CHEMISTRY RESEARCH AND APPLICATIONS

INTERACTIONS OF AQUEOUS-ORGANIC MIXTURES WITH CELLULOSE

CHEMISTRY RESEARCH AND APPLICATIONS

Additional books in this series can be found on Nova's website under the Series tab.

Additional E-books in this series can be found on Nova's website under the E-books tab.

CHEMISTRY RESEARCH AND APPLICATIONS

INTERACTIONS OF AQUEOUS-ORGANIC MIXTURES WITH CELLULOSE

M.I. VORONOVA
O.V. SUROV
AND
A.G. ZAKHAROV

Novinka
Nova Science Publishers, Inc.
New York

Copyright © 2010 by Nova Science Publishers, Inc.

All rights reserved. No part of this book may be reproduced, stored in a retrieval system or transmitted in any form or by any means: electronic, electrostatic, magnetic, tape, mechanical photocopying, recording or otherwise without the written permission of the Publisher.

For permission to use material from this book please contact us:
Telephone 631-231-7269; Fax 631-231-8175
Web Site: http://www.novapublishers.com

NOTICE TO THE READER

The Publisher has taken reasonable care in the preparation of this book, but makes no expressed or implied warranty of any kind and assumes no responsibility for any errors or omissions. No liability is assumed for incidental or consequential damages in connection with or arising out of information contained in this book. The Publisher shall not be liable for any special, consequential, or exemplary damages resulting, in whole or in part, from the readers' use of, or reliance upon, this material.

Independent verification should be sought for any data, advice or recommendations contained in this book. In addition, no responsibility is assumed by the publisher for any injury and/or damage to persons or property arising from any methods, products, instructions, ideas or otherwise contained in this publication.

This publication is designed to provide accurate and authoritative information with regard to the subject matter covered herein. It is sold with the clear understanding that the Publisher is not engaged in rendering legal or any other professional services. If legal or any other expert assistance is required, the services of a competent person should be sought. FROM A DECLARATION OF PARTICIPANTS JOINTLY ADOPTED BY A COMMITTEE OF THE AMERICAN BAR ASSOCIATION AND A COMMITTEE OF PUBLISHERS.

LIBRARY OF CONGRESS CATALOGING-IN-PUBLICATION DATA

Available upon Request
ISBN: 978-1-61668-766-3

Published by Nova Science Publishers, Inc. ✟ New York

Contents

Preface		vii
Introduction		1
Chapter 1	Application of Effusion Technique for Sorption Study on Cellulose	3
Chapter 2	Interactions of Water-DMSO Mixtures with Cellulose	23
Chapter 3	Adsorption of Aqueous-Organic Mixture Components on Cellulose	39
Chapter 4	Adsorption of Phenol and Toluene from the Gas Phase and Aqueous Solutions on Cellulose	51
Chapter 5	Sorption of Phenol on Cellulose from Binary Aqueous-Organic Mixtures	59
References		67
Index		71

PREFACE

Water-dimethylsulfoxide (DMSO)-cellulose and water-acetonitrile (AN)-cellulose systems are studied. The sorption isotherms of the components on cellulose from water-organic mixtures are obtained, immersion heats of cellulose in water-DMSO and water-AN mixtures and heats of phenol dissolution in the above-mentioned mixtures are measured. Based on analysis of literature and experimental data an assumption is made, that the solvation features of phenol are due to the features of intermolecular interactions in binary water-DMSO and water-AN mixtures. The thermodynamic characteristics of phenol dissolution as well as sorption character on cellulose change drastic at concentrations corresponding to change of character of cluster formation in water-DMSO and water-AN mixtures. It is shown that sorption of the components of binary water-organic mixtures and phenol sorption from these mixtures on cellulose depend on the intermolecular interactions in the water-DMSO and water-AN solutions.

INTRODUCTION

Cellulose processing technology based on formation of cellulose ethers includes processes that take place or begin in heterogeneous medium. Solid-phase reactions are limited by the ratio of the diffusion rates of the reagents and the chemical reaction of the hydroxyl groups in cellulose. The reactivity of the hydroxyl groups in conducting reactions in the solid phase is determined by thermodynamic and to a significant degree, kinetic factors, i.e., the equilibrium flexibility and segmental mobility of the molecular chain in the solid phase. All of these characteristics are a function of the molecular, supramolecular, and morphological structure of cellulose. The participation of hydroxyl groups in formation of system of infra- and intermolecular bonds, which greatly determines the physicochemical properties of cellulose, is a very important factor. Solvation processes have a great effect on the equilibrium flexibility and segmental mobility of the molecular chain. Studying the effect of liquid media on polymeric materials will produce information on the features of its structure, in particular, the character and intensity of intermolecular interactions.

The interactions of cellulose with low molecular weight liquids affect greatly the processes of cellulose treatment using both chemical and physical methods. Most processes for the isolation of cellulose from plant materials and physicochemical cellulose processing are based on the interaction of cellulose with low molecular weight liquids. In spite of a large number of works concerned with the sorption properties of cellulose, the question of the mechanism of its interaction with low molecular weight liquids remains open. To gain insight into this problem, we must determine the physicochemical characteristics of the interaction of the polymer with aqueous–organic mixture components and their relation to the structure of solutions. Sorption from solutions on polymeric materials forms the basis of many physicochemical processes. In particular, these are processes associated with the vital activity of a living body, which depends on the capability to sorb through respiratory and digestive organs and through skin various substances that stimulate or prevent normal functioning of the body. Industrial processes also involve accumulation of certain compounds and utilization of by-products. Polymeric sorbents exhibit certain specific features governed by their properties and structure. On the molecular level, these materials are mixtures of macromolecules of different lengths and irregular structure. The supramolecular structure of the polymers is characterized by the existence of areas significantly differing in the extent of macromolecular ordering. To a first approximation, sorbing amorphous regions and sorption-inactive crystallites can be distinguished. Despite a large number of papers dealing with sorption properties of polymers, the mechanism of the interaction of polymers with mixtures of low molecular weight liquids (solutions or emulsions) is still poorly understood. Study and prediction of the properties of systems consisting of a polymer and an aqueous–organic solvent is necessary for solving problems related, in particular, to the development of sorbents based on composites of natural and synthetic polymers for water treatment to remove aromatic hydrocarbons and petroleum products. To understand the nature of the interaction of low molecular weight liquids with a polymer, it is necessary to elucidate the physicochemical features of the interaction of a polymer with components of an aqueous–organic solution and the correlation of these features with the solution structure. The solvation effect shows a nonlinear dependence on the component ratio, owing to the fact that the solvation shell around a solute molecule or around functional groups of a polymer differs in the composition from the bulk of the binary solvent. Deviations from a thermodynamically ideal binary system arise when molecules of water and organic solvent are mixed nonhomogeneously on the level of clusters. Hence, in such nonideal binary systems the preferential solvation occurs more readily. The molecular cluster formation in binary solvents determines the solvation of solutes. This preferential solvation may be a key moment in sorption on a polymer.

Chapter 1

APPLICATION OF EFFUSION TECHNIQUE FOR SORPTION STUDY ON CELLULOSE

Currently a great body of experimental data has been gained on the interaction of cellulose with water [1-5]. It is commonly accepted that only amorphous domains in the cellulose structure are involved in this interaction. Furthermore, the thermodynamic state of sorbed water differs from that in the bulk of the liquid, which is reflected in, e.g., decrease in the diffusion coefficient, spin-spin relaxation time, and freezing temperature [6]. This difference was also confirmed by the model calculations of the structure and properties of water in thin layers and small volumes [7]. It was demonstrated that the change in the characteristics of water in limited volumes is caused primarily by rearrangement of the hydrogen bonds. In small pores and capillaries (<50 Å in diameter), water cannot take the structure typical of bulk liquid water. The computer simulation data [7] show that, in thin films with a thickness well comparable with the diameter of the water molecules, disjoining pressure oscillations occur, playing the role of a hydration pump. Cellulose-bound water may considerably influence the properties of cellulose-based materials, e.g., medical products. Therefore, it is important to know how various methods of mechanical treatment (crushing) influence the sorption properties of cellulose, what amount of water is sorbed or desorbed at various relative humidity and temperatures, and, at last, what is the thermodynamic state of cellulose-bound water in relation to the moisture content.

The subject of the investigation was powdered cellulose with the degree of polymerization equal 700 and the degree of crystallinity 58%. Two kinds of cellulose samples were investigated. The first one swelled in contact with liquid water up to 20.0% equilibrium humidity and the second kind of samples was moistened by water vapor up to equilibrium humidity equal 6.5%.

A cellulose sample was placed into the effusion cell (figure 1) in a glass container. The design of the experimental cell provides a device for effusion orifice closing during establishing the steady regime of measurements. The experimental cell was made of stainless steel with internal volume of 3.5 mm^3 and effusion orifice diameter 0.048 mm. The internal diameter of the glass container is about 8 mm; the ratio of sample surface area to effusion orifice area is about $3 \cdot 10^4$. Mass of the sample was about 100 mg.

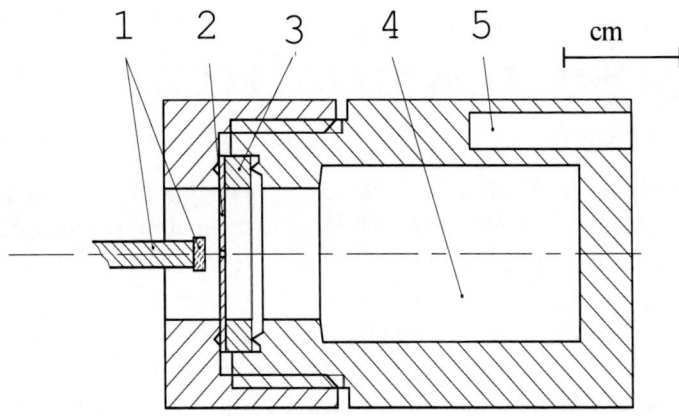

Figure 1. Design of the effusion cell for investigation of absorption equilibrium. Component parts of the experimental cell are: components of capsulation (1) of the effusion orifice, the diaphragm (2) with the effusion orifice, the vacuum seal (3) of the diaphragm, the internal chamber (4) for sample placement, the hole for thermocouple placement (5) into the cell wall.

The water vapor flow through the effusion orifice was measured using a MI-1201 mass-spectrometer that was specially equipped for the effusion experiments [8]. The ion-accelerating device maintained 5.0 kV voltage. Two techniques were applied to reduce the background of ion flow. First, the cell with the heater was placed into the nitrogen trap with the individual evacuation system. Second, the analytical part of the mass-spectrometer was heated during 30 minutes before the experiments. The ratio of the background to a useful signal was estimated using a molecular beam shutter. The background part was about 5% of the measurable flow for high-moistened samples and 80% of flow for the samples with humidity less than 1%. It should be noted that in the latter case effusion of water vapors from the cell is insignificant and does not affect the background value and that it can be easily taken into consideration. Temperature was maintained using two tungsten-rhenium VR 5/20 thermocouples. One of them was placed into the heater area; the other was assembled in the wall of the cell. The accuracy of temperature maintenance is within 3 K.

The sample looses mass within of 0.1-0.001 mg depending on the cellulose humidity when the effusion orifice opens for 0.5-1.5 minutes to measure water vapor flow. The intensity of H_2O^+ ion gas current changed less than 0.7% during the measurement period. All measurements were conducted for fixed humidity of the samples. A fixed amount of water is evaporated and then the effusion cell is in closed condition up to 20 hours in order to maintain the equilibrium of vapor pressure measurements. The pressure of water vapor over samples was determined for cellulose humidity from 0.38% to 20%. The measurements were carried out for the experimental points with the temperature increasing and decreasing in the range from 278 K to 303 K. Pressure stabilizes much later than temperature and thus the experimental cell was kept at constant temperature during 90-120 minutes before successive mass spectrum registration. Therefore, only equilibrium experimental points were taken into account. The experimental values of H_2O^+ ion current I are shown in figure 2 as a function of temperature for different samples humidity.

The pressure value P in the experimental cell was calculated as a product of the gas current I and temperature T using Equation (1).

$$P = kIT, \qquad (1)$$

where k is a constant of sensitivity of mass-spectrometer for the measured compound. If the sample looses the mass dm during the period $d\tau$ because of water evaporation through the effusion orifice, the value P can be expressed using the Hertz-Knudsen equation:

$$P = \frac{dm}{S_{effective} d\tau} \sqrt{\frac{2\pi RT}{M}}, \qquad (2)$$

where $S_{effective}$ is the effective area of an effusion orifice, M is the molecular mass of a compound, and R is the gas constant. The sensitivity constant k was obtained from Equations (1) and (2) by integration over the time $d\tau$.

$$k = \frac{\Delta m \sqrt{2\pi R/M}}{S_{effective}} \left(\int_0^\tau I\sqrt{T} d\tau \right)^{-1} \qquad (3)$$

A similar method for pressure calculation was proposed by the authors [9]. The time τ for the open cell state during which effusion tooks place was registered to calculate the integral in Equation (3).

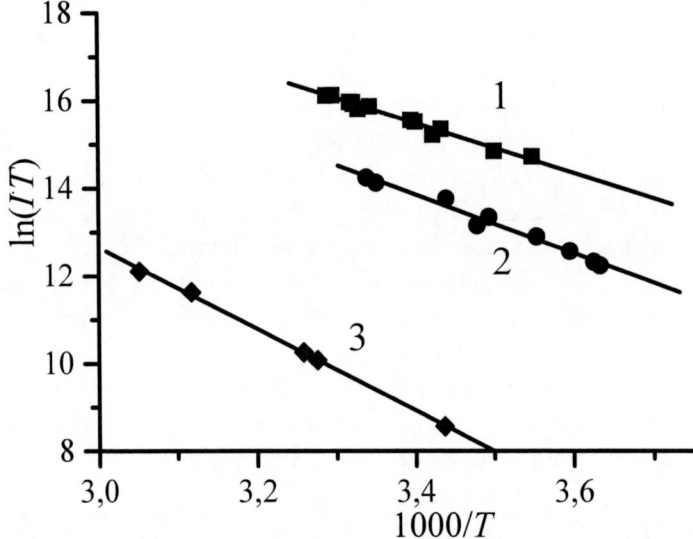

Figure 2. The experimental values of the H_2O^+ ion current I at different temperatures for sample humidity of 12.3% (1), 5.6% (2), and 0.98% (3).

The humidity value a_i can be evaluated as a function of time τ_i by the following transformation of Equation (3):

$$a_i = \left(\int_0^{\tau_0} I\sqrt{T} d\tau - \int_0^{\tau_i} I\sqrt{T} d\tau \right) k_m / m_C^0 , \qquad (4)$$

where τ_0 is the total evaporation time, m_C^0 is the mass of dry cellulose, and the constant $k_m = kS_{effective}/\sqrt{2\pi R/M}$ can be calculated by weighing the initial sample and the sample dried in the effusion cell at 100 °C and decreasing the pressure of the water vapor to 0.2 Pa:

$$k_m = m^0_{H_2O} \Big/ \int_0^\tau I\sqrt{T}d\tau ,\qquad(5)$$

where $m^0_{H_2O}$ is the mass of the evaporated water. The calculated $I\sqrt{T}$ values and humidity a as functions of evaporation time τ are shown in figure 3. Numerical integration in Equations (4) and (5) was made by standard methods.

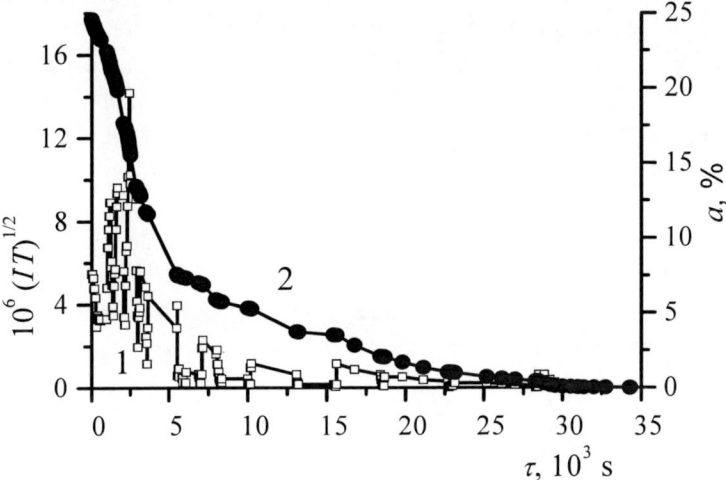

Figure 3. The $I\sqrt{T}$ values (1) and humidity a (2) calculated from the experimental data.

The values of water vapor pressure in the experimental cell were obtained in the range from 0.6 Pa to 4000 Pa. The molecular regime of the effusion is observed only at low humidity where pressure is less than 6 Pa. Then the free path of a molecule in vapor is about 0.2 mm, i.e. it is greater than the effusion orifice diameter. A transient behavior of the effusion should be taken into account at higher pressure.

Evaporation of alkali metals salts has been investigated [10] by the effusion methods both for the molecular and gas-dynamics regime of the effusion. The vapor pressure values were calculated from mass loss of a sample. An overvaluation of the pressure calculated for the gas-dynamics regime is compared to the molecular regime of the effusion for the compounds studied. A correction factor was found to be 1.5 and it was the same for other compounds [10].

The pressure was determined on the base of vapor current flowing from the effusion cell which goes into the ionization chamber. If the wall of the effusion orifice is thin and the effusion has molecular behavior, the orientation distribution of the vapor stream follows the cosine rule [11]. At higher pressure, the molecular collisions are observed in the stream that has left the effusion cell. It leads to violation of the cosine distribution. In that case, calculation of the pressure value from the ion current value I using Equation (1) gives low value compared to the pressure value in the experimental cell.

A reliable model of pressure calculation should consider the angle distribution of the effused molecules and their collisions in the molecular beam. Due to the lacking of such model at present, a correction for the specific effusion regime was determined empirically. We took into account the experimental fact, that the vapor pressure above different kinds of cellulose with 25% humidity is the same and equals to 0.86±0.04 from the pressure of saturated water vapor $P_{saturated}$ [12]. The correction for the specific effusion regime was determined by comparing that value with the calculated pressure value (Equation (1)). This correction decreases nearly exponentially with pressure decreasing. Therefore, pressure of saturated vapor can be calculated using the following approximation:

$$P_{saturated} = P[1.6 - 0.6\exp(-P/565)]. \qquad (6)$$

Finally, after the correction was taken into account, the pressure dependence on temperature was approximated by Clapeyron-Clausius equation:

$$\ln(P) = -a/t + b, \qquad (7)$$

where $a = \Delta H_T^o / R$ and $b = \Delta S_T^o / R$.

Therefore, ΔH_T^o enthalpy and ΔS_T^o entropy values of the total process including desorption and evaporation of water can be calculated from the data obtained using the second law of thermodynamics. The values of enthalpy and entropy of water evaporation are the reference values [13] and thus thermodynamics properties of the water desorption process from cellulose materials can be found. The correction for the specific effusion regime does not influence drastically the values of enthalpy and entropy of desorption. For instance, it gives 5 kJ/mole higher value of the desorption enthalpy at humidity of 20% compared with the value calculated without the use of Equation (6). The correction value becomes less than the experimental error at lower humidity and at lower pressure.

The differential enthalpy and entropy of the water desorption process from cellulose are shown as functions of humidity in figures 4 and 5, respectively.

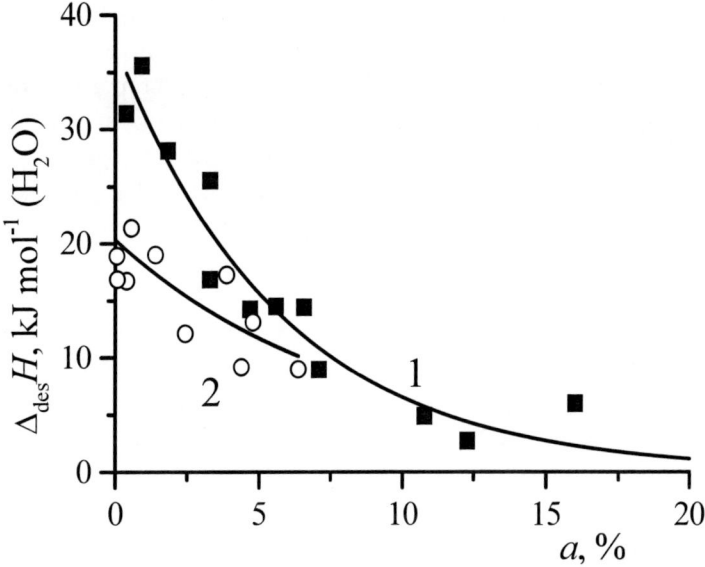

Figure 4. Enthalpy of water desorption as a function of cellulose humidity for the samples preliminarily swelled contacting with liquid water (1) and kept in water vapor (2).

One can see the significant difference of the differential enthalpy of water desorption from cellulose samples which were swelled contacting with liquid water up to humidity 20.0% and the samples that were moistened by water vapor up to equilibrium humidity of 6.5%. That difference can be explained by distinctions of the sorption mechanism in the cellulose - liquid water and cellulose - water vapor systems.

Small inclusions i.e. clusters are formed by water in contact of dry cellulose with liquid water [14]. These water clusters become stabilized due to the grouping of the polymer chains and formation of micropores around the clusters. It results in the decrease of the entropy value of the surrounding polymer chains. It influences the other polymer chains; their intermolecular interactions decrease and, consequently, macromolecular order is destroyed. These effects grow because the distortion of the order leads to the formation of additional cluster inclusions and to further developing a micropores structure of the polymer. Simultaneously, water clusters become enlarged by means of their coalescence or fusion. The total surface of the water clusters and their surface energy become smaller. If the molecular frame of the polymer is pliable enough, the cluster enlargement process develops to formation of slit-like pores that are favorable energetically. Thus, one can accept the proposal that the micropore structure formation is accompanied by disappearance of the phase interface and that the water – cellulose system becomes one phase. Therefore, the physical sorption phenomena on such micropore polymers are adequate to bulk dissolution in some respects. Then the sorption thermodynamics should be described by means of the Gibbs-Dugem equation instead of the Gibbs equation. The chemical potential of cellulose is changed in the water absorption just as in the solution formation.

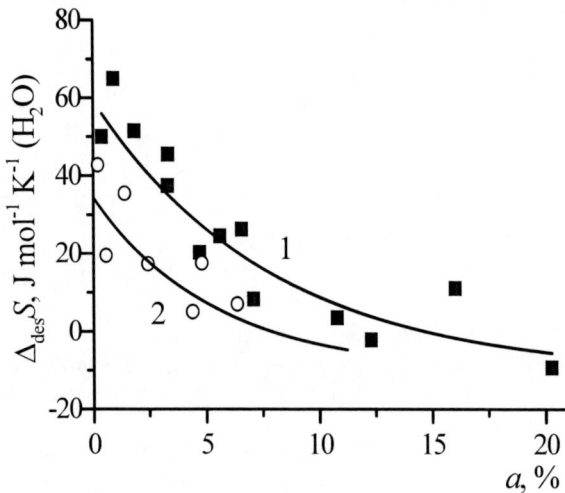

Figure 5. Entropy of water desorption as a function of cellulose humidity for the samples preliminarily swelled contacting with liquid water (1) and kept in water vapor (2).

Thus, cellulose accumulates a significant amount of water in the slit-like pores in liquid water adsorption. According to work [15], cellulose has been impregnated with water for a few seconds. In contrast, the desorption process lasts a few hours. That is why the meniscuses with negative curvature can arise in the slit-like pores. Such formations are characterized by lower vapor pressure compared to the plain interface. The curvatures of the meniscuses increase with the desorption process and the new water layers are included into that process and, consequently, the difference of enthalpy of water desorption becomes larger.

No water clusters are formed in cellulose when the cellulose sample is placed into water vapor and kept there up to a humidity of 6.5%. Water is adsorbed in the most active adsorption centers. Cellulose does not swell in volume; the micropore structure does not arise, and the chemical potential of cellulose changes slightly. As a result, the value of vapor pressure is higher and the value of water desorption enthalpy is lower for cellulose samples moistened in water vapor than in liquid water (figure 6).

The enthalpy values of water desorption at 291 K extrapolated to zero humidity are $\Delta_{des}H=23\pm12$ kJ mol^{-1} and $\Delta_{des}H=37\pm12$ kJ mol^{-1}, respectively for both types of moistening. A lower value of the entropy desorption $\Delta_{des}S=36\pm19$ Jmol^{-1} K^{-1} for the sample moistened by water vapor compared to $\Delta_{des}S=59\pm19$ Jmol^{-1} K^{-1} for samples moistened by liquid water confirms the formation of micropores structures.

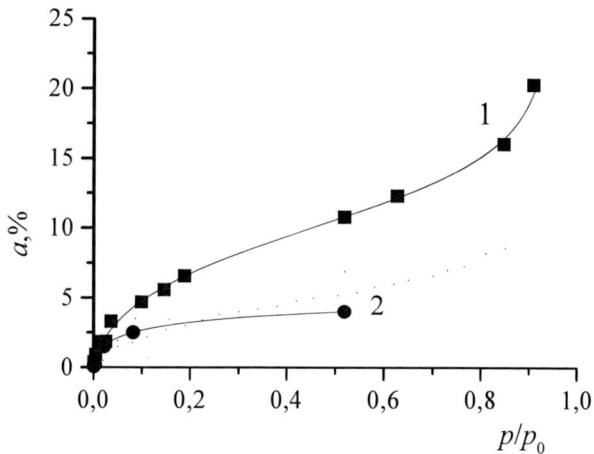

Figure 6. Isotherms of water desorption for cellulose samples preliminarily swelled contacting with liquid water (1) and kept in water vapor (2).

Thus, it is experimentally confirmed that the mechanism of water sorption by cellulose is different for cellulose in contact with liquid water and water vapor to humidity of 6.5%. The increase of sample humidity of cellulose in contact with water vapor seems to result in the disappearance of distinctions in sorption mechanism.

The method proposed allows the humidity measuring without sample extraction from the experimental cell and its subsequent weighing. The inaccuracy value of the humidity measurements is permanent over the whole experimental range. That value depends on the precision of the measurements of the ion currents using a mass spectrometer. Monitoring the equipment stability during one year showed that the fluctuation value of sensitivity is within 10%.

Further, desorption of water from mechanically and chemically treated cellulose was studied using the effusion technique. Sulfite cellulose was mechanically treated on an extruder or shock-reset machine. The chemical treatment involved hydrolysis of cellulose for 2 h at 365 K in 10% HNO_3. As known, using the effusion technique, desorption isotherms of solvents can be obtained over the entire range of the relative vapor pressure at low time consumption.

Determination of the vapor pressure by the Knudsen's effusion technique is based on measuring the vapor effusion rate from a cell to a vacuum through an orifice in the membrane [11]. A great diversity in design characteristics of the existing effusion installations originates from a wide spectrum of research objectives. In our case, the design criteria were maximal simplicity and availability of equipment in combination with high precision and reproducibility of results. The experimental effusion installation was designed on the platform of a VOU-1A vacuum unit providing the residual pressure in the system no higher than 1×10^{-4} mm Hg. A cylindrical effusion cell (4 cm^3 in volume) was placed in an electric furnace. The temperature of the cell was controlled to within 0.1 K using a thermocouple assembly and a TRM-34 doublecircuit thermoregulator. Additionally, the effusion cell was equipped with an electromagnetic valve hermetically closing the effusion orifice until the desired temperature was attained. A cellulose sample (~100 mg), vacuum-dried at 338-343 K to constant weight, was wetted in a glass ampule with a small portion of distilled water (on the basis of 500 mg of water per gram of cellulose) and placed in the effusion cell. The cell was hermetically sealed, weighed, and allowed to stand in an air bath oven at the temperature of the planned effusion experiment for at least 24 h. The cell was placed in a furnace heated to the desired temperature; the system was evacuated and allowed to stand at the preset temperature for at least 120 min. Then the orifice was opened, and the effusion experiment started. After completion of the experiment, the orifice was closed; the system was filled with air, and the ampule with the sample was weighed again. From the weight loss, effusion time, and temperature, we estimated the vapor pressure by the Knudsen equation (8):

$$P_k = (\Delta m / \alpha \beta S_h \tau)(2\pi RT/M)^{1/2}, \tag{8}$$

where Δm is the weight loss through effusion; α is the condensation coefficient; S_h is the orifice area; β is the Clausing coefficient characterizing the resistance to mass transfer; τ is the effusion time; R is the gas constant; T is the temperature; and M is the molecular weight of the substance.

Following [16, 17], we used naphthalene and benzoic acid as reference compounds. The requirements to reference compounds for calorimetry of vaporization (availability, usability, high purity, and safety) essentially coincide with those to reference compounds for calorimetry of other processes such as combustion and dissolution. However, in calorimetric study of vaporization, it is needed to have quite a set of such compounds with the vapor pressures ranging over a broad interval, insofar as the fact that the method provides accurate results in a certain temperature range by no means guarantees the same accuracy over another temperature range. Many researchers prefer naphthalene, because, like other aromatic compounds, it can be quite easily purified using such methods as recrystallization, sublimation, zone melting, directional crystallization, and chromatography. Using these methods allows reduction of foreign impurities, e.g., anthracene, by several orders of magnitude. Benzoic acid is also often used as a reference. It was recommended by IUPAC, as having well-defined and reproducible melting and evaporation points. In this study, benzoic acid (mp 395.5 K) and naphthalene (calorimetrically pure grade, mp 353.43 K) were further purified using fractional sublimation in a high vacuum. The middle fraction was used for making calibrations. Calibration with naphthalene was carried out over the temperature range from 298 to 318 K at an effective orifice area ($S_{eff} = \beta S_h$) of 2.21×10^{-7} and 2.7×10^{-8} m^2; and by benzoic acid, from 323 to 348 K at S_{eff} 8.48×10^{-7} and 2.21×10^{-7} m^2. We used a specially designed tool, which allowed making regularly round-shaped effusion holes up to 0.01 mm in diameter in 0.1-mm Al foil. The orifice areas were measured microscopically at a magnification of 200. The Clausing coefficient β was estimated as recommended in [11]. The experimental heats of sublimation of naphthalene and benzoic acid (72.4±0.6 and 89.8±0.7 kJ mol^{-1}, respectively) were well consistent with the recommended values [16, 17]. Despite the advantage of the effusion technique, as a time-saving method, over, e.g., the isopiestic method, it has some limitations. It is well known that the applicability of the effusion method is limited to a pressure of 10 Pa and an effusion orifice radius of 1 mm. This is caused by the fact that the condition of molecular flow is met if the free path of the molecules in the effusion cell significantly exceeds the orifice diameter. However, there are examples in the literature for the application of the effusion technique to measuring pressures considerably exceeding 10 Pa. In these cases, various corrections should be introduced to take into account the deviation from the molecular flow mode. Furthermore, it was demonstrated in [11, 16] that, at a sufficiently small diameter of the effusion hole, all the limitations of the method, originating from the effusion mode deviation from the molecular flow, are eliminated; the condensation coefficient α becomes equal to unity, and the measured vapor pressure no longer depends on the effusion orifice area and can be accepted equal to the true saturated vapor pressure.

Figure 7 shows the water vapor pressure estimated by the Knudsen equation (8), as influenced by the effusion orifice size (at α = 1). As seen, with decreasing orifice area, the estimated vapor pressure approaches the true value. Therefore, for the effusion cell used in this study, the estimated water vapor pressure corresponds to the true value at a orifice diameter of about several hundredth parts of millimeter and smaller. At such a small size of the effusion orifice, the sensitivity of the method, and, therefore, its accuracy, decreases.

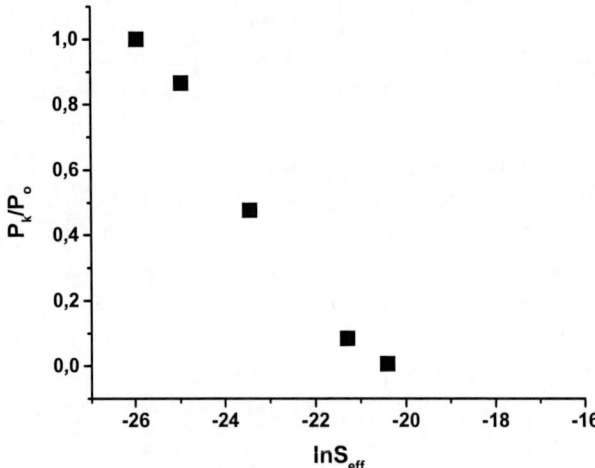

Figure 7. Water vapor pressure in the Knudsen cell as a function of the effective area of the effusion orifice ($S_{eff} = \beta S_h$) at 298 K. (P_k) Vapor pressure estimated by the Knudsen equation (8) and (P_0) saturated vapor pressure at 298 K.

Therefore, there are certain difficulties in using the effusion method for studying water desorption over a wide pressure range. First, the correction factors used to take into account the deviation from the molecular flow mode are empirical, being dependent on the investigated system and the design of the effusion cell. Second, decrease in the orifice area results in deterioration in the sensitivity and accuracy of the method. Third, as will be demonstrated below, in the system in hand, the condensation coefficient α is pressure-dependent, so that its application becomes questionable in our case. To obviate complications inherent in the effusion method in the range of high relative water pressures, employing all the advantages of the method in measuring low pressures, we acted as follows. We prepared aqueous solutions of sulfuric acid of various concentrations and measured the water vapor pressures over these solutions using the effusion method (effective orifice area 8.206×10^{-10} m^2). The results were used to construct calibration curves coupling the vapor pressure obtained by the Knudsen method P_k with the true water vapor pressure P_s corresponding to a given concentration of sulfuric acid [18]. An example of such a calibration curve, obtained at 298 K, is presented in figure 8. As seen, this is not a straight line, but a curve which can be extrapolated by a third-order polynomial with a correlation factor of 0.995. The calibration curves obtained at 303 and 308 K are similar. Even at low pressures, the curve deviates from linearity, and, as the pressure increases, the function P_s vs. P_k becomes much more complex, which can be attributed to alteration in the vapor flow mode from the effusion orifice.

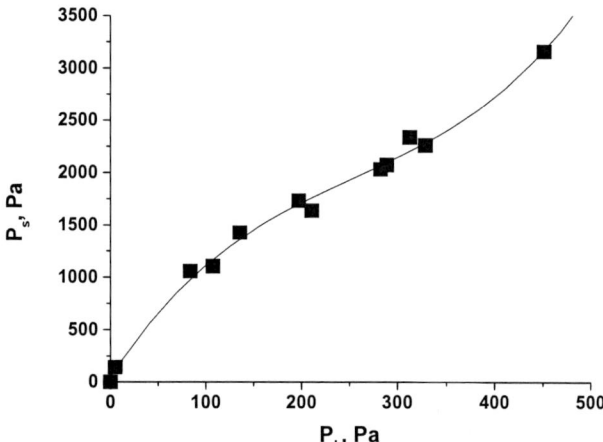

Figure 8. Calibration curve for determination of the true water vapor pressure P_s from the experimental P_k values obtained using the effusion technique at 298 K.

Using the above procedure, we obtained desorption isotherms of water from the cellulose samples at 298, 303, and 308 K. The water vapor pressure was determined to within 10%. Figure 9 shows the primary data obtained in the effusion experiment and then used for construction of the desorption isotherms. In the experiment, we determined the water vapor pressure over a cellulose sample wetted with water. The moisture content in cellulose, at which the equilibrium vapor pressure P started to deviate from the saturated vapor pressure P_0 at a given temperature (in other words, when P/P_0 started to deviate from 1), corresponded to the maximal sorption of water on cellulose.

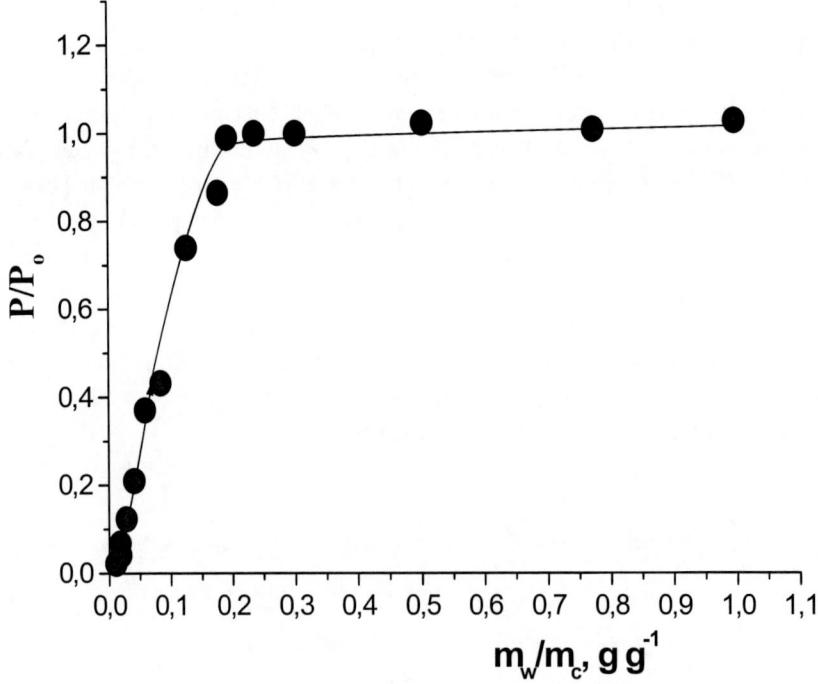

Figure 9. Primary data obtained in the effusion experiment on desorption of water from cellulose sample I at 298 K. (m_w/m_c) Water content in cellulose (g/g) and (P/P_0) relative vapor pressure of water.

Figure 10 shows the water desorption isotherms from cellulose treated mechanically using an extruder (sample I) or shock-reset machine (II), or chemically with dilute HNO_3 (III). For comparison, figure 11 shows the water desorption isotherms from sample I, obtained at 298 K using both the effusion and isopiestic (sample is kept in a desiccator at a fixed humidity) methods.

Application of Effusion Technique for Sorption Study on Cellulose 17

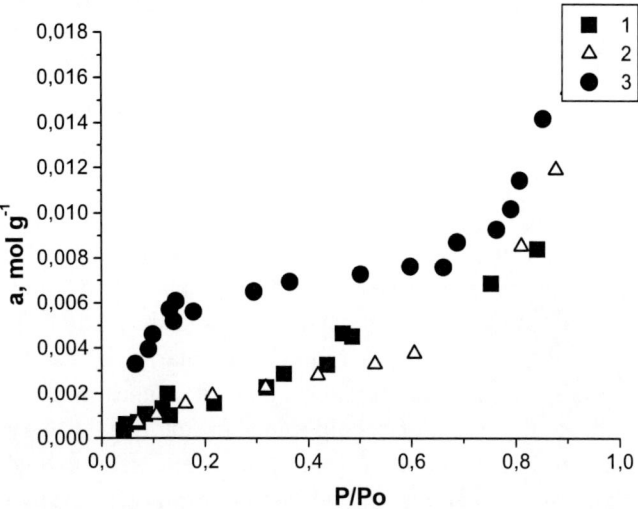

Figure 10. Water desorption isotherms from cellulose samples (*1*) I, (*2*) II, and (*3*) III at 298 K. (a, mol g^{-1}) Equilibrium sorption of water on cellulose and (P/P_0) relative vapor pressure of water; the same for figure 11.

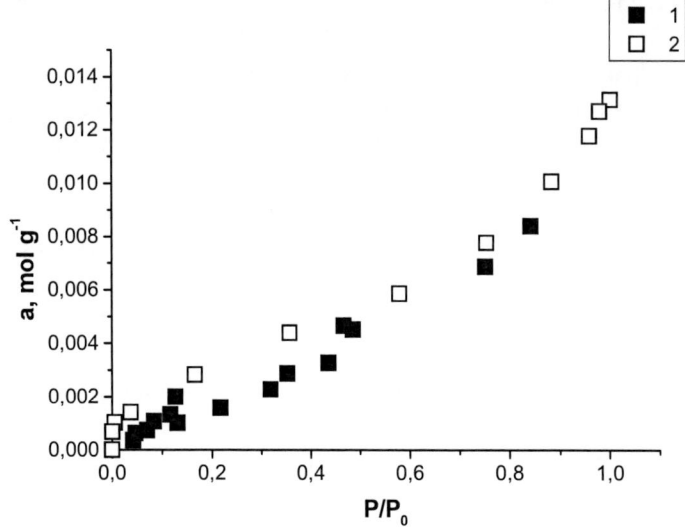

Figure 11. Water desorption isotherms from cellulose sample I, obtained by (*1*) effusion and (*2*) isopiestic methods.

The isopiestic experiments were carried out over the relative vapor pressure range P/P_0 = 1- 0.0003. A cellulose sample of about 100 mg, vacuum-dried at 338-343 K to constant weight, was brought into the isopiestic equilibrium with aqueous sulfuric acid of a fixed concentration, for which the water vapor pressure over the solution is known [18]. The experiment was carried out in an air bath oven. The temperature was controlled to within 0.1 K. The equilibration time was from 7 to 21 days. The amount of sorbed water was determined gravimetrically. The same method was used to control the equilibrium conditions in the system. The results obtained by both methods are in satisfactory agreement, even though the isopiestic method gives slightly overestimated values for water sorption, particularly, in the range of low relative pressures, which is probably due to longer time required for establishment of the sorption equilibrium in the isopiestic experiments. Previously we studied the equilibrium in the system water-cellulose using an MI-1201 mass spectrometer modified for performing effusion experiments [19]. Data on the H_2O^+ intensity in the mass spectrum suggested that, in the effusion experiment, the time required for establishment of the equilibrium is 90 -120 min, depending on the moisture content and temperature of cellulose samples. The dependence of the relative water vapor pressure at a given sorption value on the reciprocal temperature was then used to estimate the thermodynamic parameters of water desorption by the Clapeyron-Clausius equation (7)

The enthalpies and entropies of desorption of water from the cellulose samples are presented in figures 12a and 12b.

The heat effects of water desorption from cellulose samples II and III at low sorption values are about 10-20 kJ mol^{-1}, decreasing to zero with increasing sorption, which is typical of the cellulose-water system. In this case, the entropy of the system is practically independent of the amount of sorbed water, differing only slightly from zero, suggesting that the interaction in the system is accompanied by ordering of the water molecules and disordering of the cellulose segments. Sample I stands out against a background of the other samples. For sample I, the enthalpy of water desorption at low sorption is about 55 kJ mol^{-1}, and the entropy of desorption changes in parallel to the enthalpy, i.e., the thermodynamic compensation effect takes place here (figure 13).

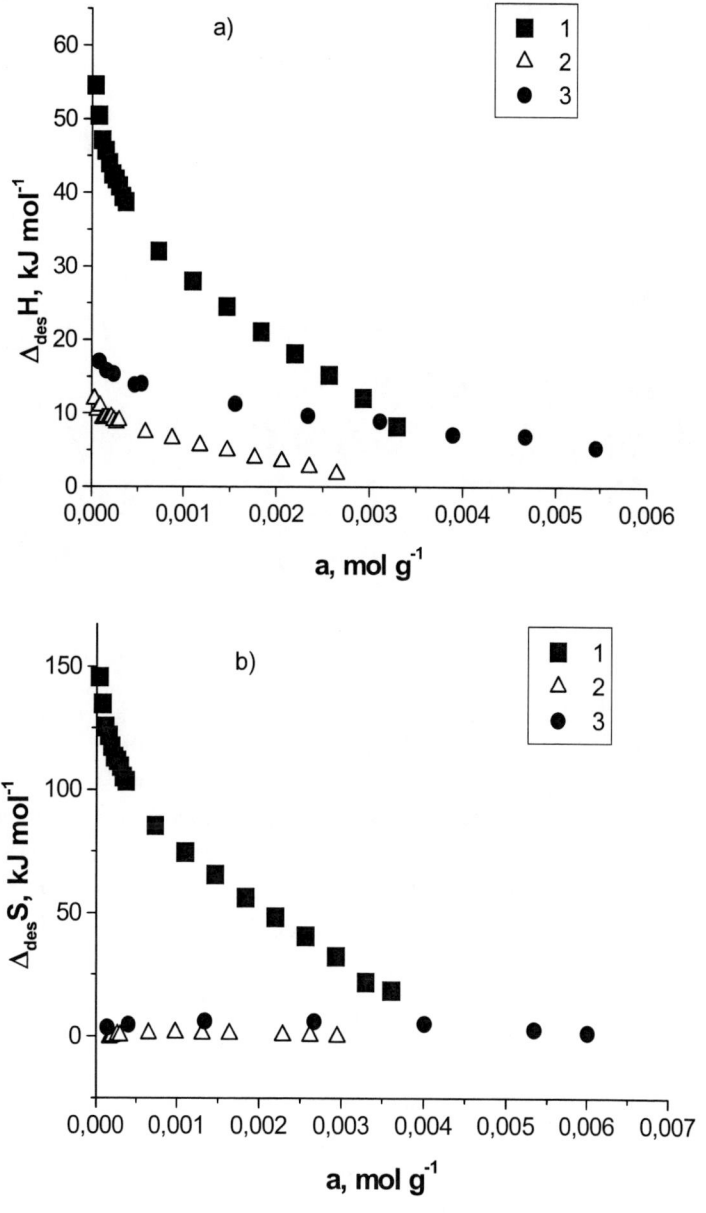

Figure 12. (a) Enthalpy and (b) entropy of water desorption from cellulose samples (*1*) I, (*2*) II, and (*3*) III. (ΔH_{des}) Enthalpy and (ΔS_{des}) entropy of water desorption from cellulose, and (*a*) equilibrium sorption of water on cellulose.

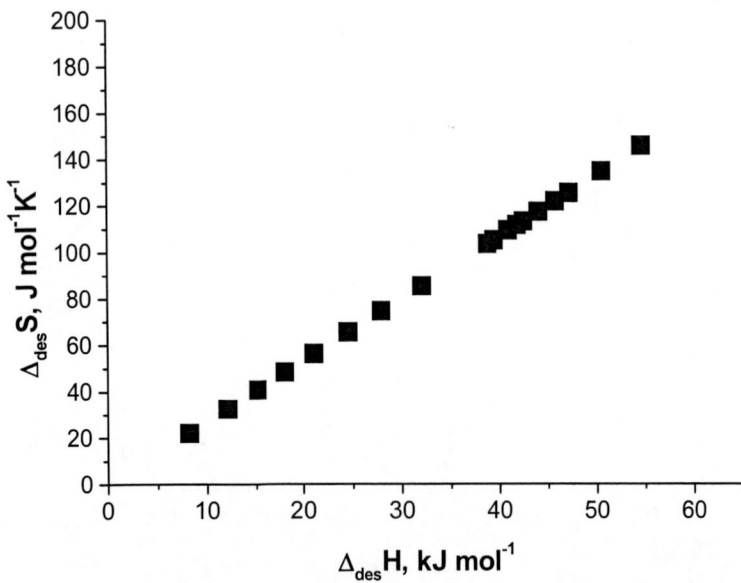

Figure 13. Thermodynamic compensation effect in water desorption from cellulose sample I. (ΔH_{des}) Enthalpy and (ΔS_{des}) entropy of water desorption from cellulose.

The compensation or isokinetic theory is widely used in physicochemical research. According to this theory, the enthalpy-entropy compensation originates from changing nature of the solute-solvent interaction due to a specific reaction, the relationship between the enthalpy and entropy being linear. Aguerre [20] applied this approach to the study of sorption of water on foodstuffs. He derived an equation relating the equilibrium sorption of water to its activity and the temperature, and demonstrated a linear correlation to occur between the enthalpy and entropy for a series of foodstuffs. It was demonstrated also that the penetration of water across the grape peel is a rate-controlling stage of drying, i.e., the resistance to water loss is provided primarily by the cell membrane [21]. The water loss can be decelerated through optimization of the storage conditions (temperature and humidity). It may be suggested from the observed abnormally high heat effects of water desorption from sample I in combination with the thermodynamic compensation effect that the interaction of water with this sample differs from that with other samples. Among the reasons could be chemical interaction of water with some impurities introduced into cellulose during pretreatment, changing number of hydration centers and characteristics of the micropores; transformation of the crystalline domains, etc. By now we cannot fix the key

factor. Nevertheless, the study of the conditions and
desorption from cellulose is of crucial importance.
sorption on cellulose materials, it is necessary to take in
of the sample. There is no doubt that data on the equilibri
are significant for optimization of drying and storage
medical products. In this context, the effusion technique sho
express method for characterization of desorption of water
solvents.

Hence, with precision calibration systems, the effusion me
used for determination of the equilibrium water vapor pressure ov
The experimental isotherms of water desorption from cellulose, o
the effusion method, are well consistent with those obtained u
isopiestic technique. Determination of the equilibrium water vapor pre
various temperatures using the effusion method allows estimation
thermodynamic characteristics of water desorption from cellulose.

Chapter 2

INTERACTIONS OF WATER-DMSO MIXTURES WITH CELLULOSE

The interactions of cellulose with low molecular weight liquids affect greatly the processes of cellulose treatment using both chemical and physical methods. Such interactions in water–DMSO mixture have a particular interest because this solvent represents a very suitable medium for obtaining cellulose with necessary properties. Also DMSO has shown to demonstrate a cryoprotective effect in biological systems and it can be used as pharmaceutical agent with very low toxicity [22]. The water–DMSO system has been studied using various experimental methods such as chemical thermodynamics methods [23-25], X-ray structure analysis [26], mass spectroscopy [27, 28], NMR [29], and calorimetry [30]. According to these experiments aqueous DMSO has three concentration ranges with different mixing schemes. The conclusion about the existence of X_{DMSO}=0.28 and X_{DMSO}=0.56 concentration bounds has been made in Ref. [23,25] taking into account the partial molar values as well as their derivatives with respect to solvent composition. According to X-ray diffraction analysis of water–DMSO mixtures [26] the same boundary concentrations have been found. The composition of the forming clusters has been studied in Ref. [27] using a special weight-spectrometric method with direct isolation of clusters from a solvent mixture. In this method the clusters are isolated (or extracted) from liquid droplets injected into the vacuum chamber where the liquid droplets are fragmented into clusters through adiabatic expansion. Even though the weak interacting molecules are vaporized from the liquid droplets, the strongly interacting molecules remain in the clusters generated through the fragmentation of the liquid droplets. Accordingly, the observed cluster structure is related to the intermolecular interactions in solution. It has been shown in Ref. [27] that, at 25 °C within the $0<X_{DMSO}<0.28$ concentration range water clusters exist up to $X_{DMSO}<0.07$ and they can interact with one or two DMSO molecules. In the range of $0.07<X_{DMSO}<0.1$ the concentration of water clusters decreases sharply, and at higher DMSO concentrations ($0.1<X_{DMSO}<0.28$) it decreases slowly, whereas the concentration of DMSO clusters increases. Within the $0.07<X_{DMSO}<0.28$ concentration range, as X_{DMSO} increases, the characteristic peaks of liquid water of the radial distribution function (2.9, 4.5 and 7.1 Å) are gradually reduced and taken over by the 2.6 Å peak and the 5.6 Å intermolecular DMSO interaction. The former peak at 2.6 Å is due to the intermolecular interferences within the DMSO molecule and the nearest neighbour O–O interactions of water [26]. However, the latter stops short at 5.3Å instead of a reaching 5.6Å (the 5.6 Å peak of pure DMSO consists of number of interferences, out of which those involving the S atom make a stronger contribution due to its larger X-ray cross sections). The fact that the first intermolecular peak is at 5.3 Å indicates that the weight of shorter distance pairs increases at the expense of longer distance pairs. This means that DMSO in this composition range exists as a more tightly bound and smaller cluster than in pure DMSO. These "small" clusters consist of a number of DMSO molecules mainly held together by S=O dipoles in an antiparallel fashion. Hydrophobic CH_3-groups point outward of a "small" cluster, and, hence, DMSO works in effect as a hydrophobic solute. This conclusion is in agreement with results of Ref. [23] where the X_{DMSO} dependence of the excess partial molar volume of DMSO demonstrates a clear signature of a hydrophobic solute. In the range of $0.28<X_{DMSO}<0.56$ water clusters do not exist, whereas there is a mixture of "small" and "large" DMSO clusters (according to the terminology suggested in Ref. [26]) and their relative quantity is significantly changed. In the DMSO-rich region ($X_{DMSO}>0.56$) DMSO molecules form "large" clusters with the same local structure as in pure DMSO. Water molecules interact with DMSO clusters as single molecule, and three-dimensional structure of liquid water is completely disordered. The zero excess partial molar enthalpy and entropy of DMSO and a small constant value of water in this concentration range also prove this conclusion [23, 25]. The solvation structure of a solute depends nonlinearly on the solvent composition, exhibiting the existence of a critical value of mixing ratios where drastic changes in the microscopic solvent structure occur. The solvation structure of various solutes (2-butanol, cyclopentanol, cyclohexanol, phenol) has been investigated in Ref. [28] where nearly the same characteristics, irrespective of the nature of the solutes have been obtained. At high water concentrations ($X_W>0.9$) the solute is preferentially solvated by water clusters, whereas below $X_W\sim0.9$, the preferential solvation by DMSO clusters takes place.

The partial vapor pressures of DMSO and water in water–DMSO and cellulose–water–DMSO systems were measured by the effusion method using a MI-1201 mass-spectrometer. The experimental technique is described in detail previously. This method consists of several stages: obtaining of the ionic current versus temperature and pressure, procedure of the partial pressure determining (taking into account the non-molecular effusion regime), and calculation of the thermodynamic desorption parameters. Two techniques were applied to reduce the background of ion flow. First, the cell with the heater was placed into the liquid nitrogen trap with the individual evacuation system. Second, the analytical part of the mass-spectrometer was heated during 30 min before the experiments. The ratio of the background to a useful signal was estimated using a molecular beam shutter.

The experimental cell was kept at constant temperature during 90–120 min before successive mass spectrum registration. Therefore, only the equilibrium experimental points were taken into account. The partial pressure value P in the experimental cell is related to the measured intensity of ion peaks of the mass spectra by the Eq. (1). The vapor pressure in the Knudsen cell at temperature T is determined by Hertz-Knudsen Eq. (2). The sensitivity coefficients for water and DMSO were separately obtained from Eqs. (1) and (2) by integration over the time $d\tau$ by Eq. (3). The Δm values for each component were determined according to the original composition of a solution. The effusion time τ was registered in order to calculate the integral in Eq. (3). Substituting Eq. (3) into (1), the equation which relates the appropriate values of vapor pressures to the quantities measured in the experiment can be obtained. The value of the integral in Eq. (3) allows us to determine the weight loss of each component evaporated from the cell. Taking into consideration the original composition of a solution we were able to calculate the composition of a solution for any point of time. This approach allows us to obtain the evaporation isotherms at different temperatures.

The concentration dependences of water and DMSO partial pressures are shown in figure 14. The water–DMSO system has a negative deviation from Raoult's law for water and changing sign deviation for DMSO. Such a behaviour presumably indicates us that the intermolecular interactions in the water–DMSO system have a complex character and an average size of associates in solution changes with increasing of DMSO concentration. The values of partial pressures of water and DMSO calculated from the total vapor pressure data [23] and the values of partial pressures obtained by the direct measurements of the time-steady intensity of molecular ion peaks using a mass-spectrometer (DMSO pressure is multiplied by 20) are shown in figure 14 for comparison. The agreement is quite satisfactory except the region of $X_{DMSO} > 0.75$.

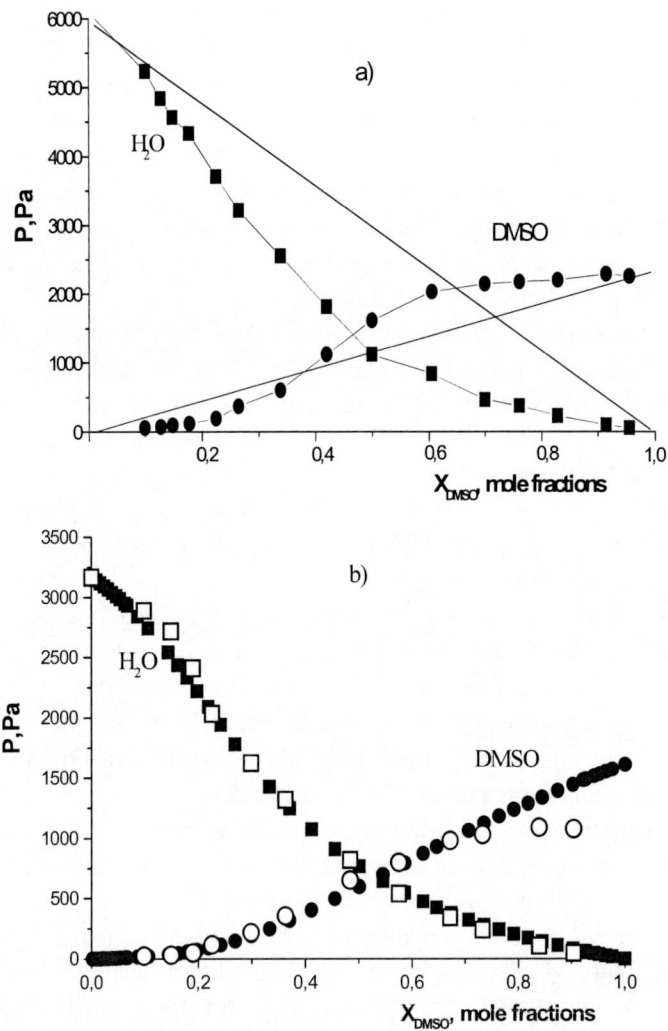

Figure 14. The partial pressures of water and DMSO (a) at temperature 308K, (b) at temperature 298K. (□) and (○)—the data from paper [23]; (■) and (●)—the present data (DMSO pressure is multiplied by 20).

In addition the partial pressures of water and DMSO for water–DMSO mixtures at temperatures 279, 303, 325, and 352 K were obtained and evaporation enthalpies for the components of a solution were calculated (see figure 15).

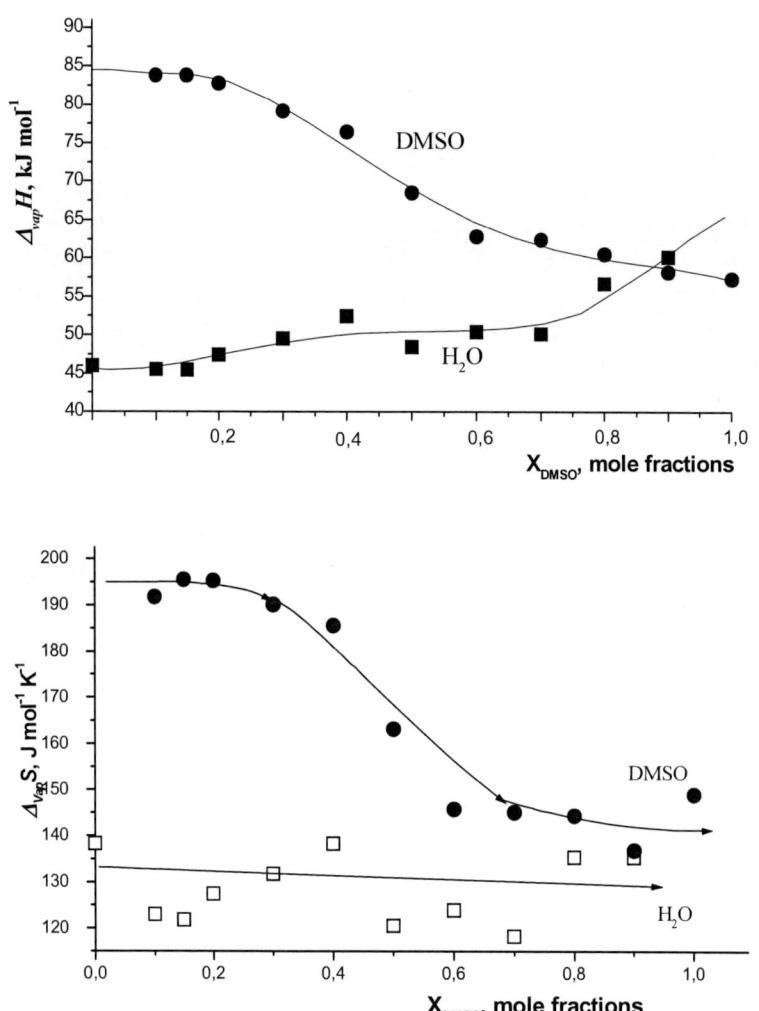

Figure 15. The evaporation enthalpies and entropies for the components of water–dimethylsulfoxide mixture (the evaporation enthalpies and entropies for pure water and DMSO are taken from [31]).

It can be seen from figure 15 that aqueous DMSO has three composition ranges with different concentration dependences of the evaporation enthalpies. Taking into consideration the data available from literature we suggest the following interpretation of mixing scheme boundaries and intermolecular interactions in these composition ranges.

In the water-rich region $0<X_{DMSO}<0.28$ there are "small" DMSO clusters and DMSO dimers with hydrophobic properties. Constant high evaporation enthalpies of DMSO are related to the hydrogen bond breakage in dimers and "small" clusters. It seems that DMSO–DMSO molecular interactions in this region are significantly stronger than those interactions in the region with high DMSO concentrations. In the water-rich region the water evaporation enthalpy is nearly constant and close to that one of pure water. Decreasing of the water mole fraction from 0.94 to 0.90 leads to considerable decreasing of water cluster formation [27] but it does not affect the values of water evaporation enthalpies. In the range of $0.28<X_{DMSO}<0.56$ the water evaporation enthalpy remains practically constant because the number of water molecules forming bonds with hydrophobic dimers and "small" clusters is small enough. Therefore the water evaporation enthalpy seems to be constant and it has nearly the same values as those observed for pure water. However there is a mixture of "small" and "large" DMSO clusters with a tendency to increase the "large"/"small" ratio with increasing of DMSO concentration. Thus, the evaporation enthalpy of DMSO decreases rapidly reaching the value observed for pure DMSO. At high DMSO concentrations ($X_{DMSO}>0.56$) the evaporation enthalpy of DMSO is practically constant and close to that one of pure DMSO because the local structure of "large" clusters is the same as in pure liquid DMSO. The increasing of water evaporation enthalpy is due to the strong interactions of single water molecules with "large" DMSO clusters. In the DMSO-rich region, DMSO molecules remain with the same molecular arrangement as in pure DMSO, whereas water structure is actually broken and water exists in the form of single molecules bonded to DMSO clusters. The evaporation entropy of DMSO decreases with increasing of the number of forming "large"DMSO clusters (see figure 15), but the evaporation entropy of water remains approximately constant.

It was shown in Ref. [27, 28], that the solvation structure of a solute is nonlinearly sensitive to the mixing ratios of water–DMSO mixtures. The factors governing the solvation structure of solutes are immediately related to the microscopic solvent structures of water–DMSO mixtures. As a consequence, the nature of the preferential solvation should be originated from the propensity that solute species dissolved in a water–DMSO solvent mixture interact with already established solvent clusters, rather than individual solvent molecules. Hence, the interactions of cellulose with water–DMSO mixtures should demonstrate a complex character. The isotherm of concentration change of the water–DMSO–cellulose system at 298 K is shown in figure 16.

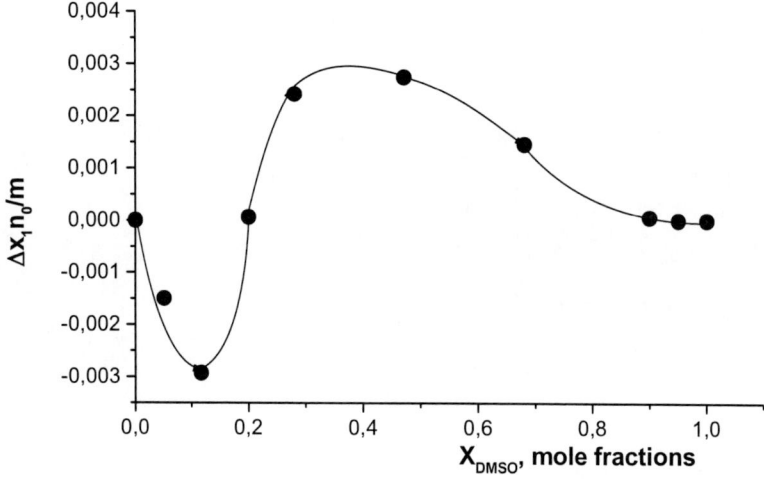

Figure 16. The isotherm of concentration change of water–DMSO mixture on cellulose.

The relation between the isotherm of concentration change and the individual adsorption isotherms can be written as

$$\frac{n_o(x_o - x)}{m} = \frac{\Delta x_1 n_o}{m} = n_1^s(1-x) - n_2^s x, \qquad (9)$$

where indexes 1 and 2 denote substances 1 and 2, respectively; x_0 is the mole fraction of 1 in the mixture before, and x is that after adsorption; n^s_1 and n^s_2 are the numbers of moles of 1 and 2, respectively, adsorbed per g. of cellulose; m is the weight (g) of cellulose sample, and n_0 is the total number of moles in the original solution. The adsorption isotherms on cellulose from water–DMSO mixtures were obtained as described in [32]. Concentrations were determined using the Pulfrich refractometer. The calibration curves were obtained for each investigated system. Differences in concentration were determined accurately by direct comparison of the solution before and after adsorption had taken place.

Taking into account the assumption made in Ref. [32] that the volume of 1 mol of ith component in an adsorption layer does not depend on its concentration in a solution and the adsorption from the solution has a displacing nature, we can derive the following formula

$$\frac{n_1^s}{(n_1^s)_m} + \frac{n_2^s}{(n_2^s)_m} = 1, \qquad (10)$$

where $(n^s{}_1)_m$ and $(n^s{}_2)_m$ are the maximal values for adsorption of components 1 and 2, respectively. The combined Eqs. (9) and (10) allow us to calculate the individual adsorption isotherms for components 1 and 2, $n^s{}_1$ and $n^s{}_2$, respectively, if the maximal values $(n^s{}_1)_m$, $(n^s{}_2)_m$ and the isotherm of concentration change $\frac{\Delta x_1 n_o}{m}$ are known.

From Eqs. (9) and (10) we have calculated values $n^s{}_1$ and $n^s{}_2$ for the water–DMSO–cellulose system (figure 17). The maximal values $(n^s{}_1)_m$ =0.011 mol of DMSO per gram of cellulose and $(n^s{}_2)_m$ =0.017 mol of water per gram of cellulose were obtained for the adsorption of the components from pure liquids as described below.

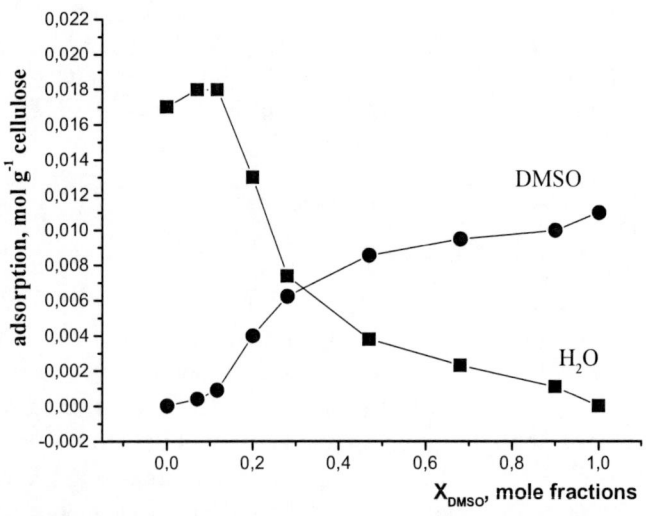

Figure 17. The calculated individual adsorption isotherms of water and DMSO from the water–DMSO mixtures on cellulose.

The unexpectedly high value of water sorption on cellulose in the presence of small amounts of DMSO was confirmed by the independent effusion experiment data.

Knudsen's effusion method is widely and successfully used to study substances with low saturated vapor pressure. However, it has, despite advantages, certain limitations: the upper applicability level of Knudsen's method does not exceed a pressure of 10 Pa, and the radius of the effusion orifice, 1 mm. This is due to the fact that the conditions of molecular outflow of a vapor from the effusion orifice are only satisfied in the case when the mean free path of molecules within the cavity of the effusion chamber substantially exceeds the diameter of the effusion orifice. When pressures substantially exceeding 10 Pa are measured, various corrections are commonly introduced to take into account deviations of the vapor outflow mode from the molecular conditions [11]. Provided that an appropriate calibration is carried out, the effusion method can be successfully used to determine the equilibrium water vapor pressure and calculate the thermodynamic characteristics of the desorption process [33].

The experimental effusion installation and the method for determining the equilibrium water vapor pressure were described in detail previously. We used a double effusion chamber schematically shown in figure 18.

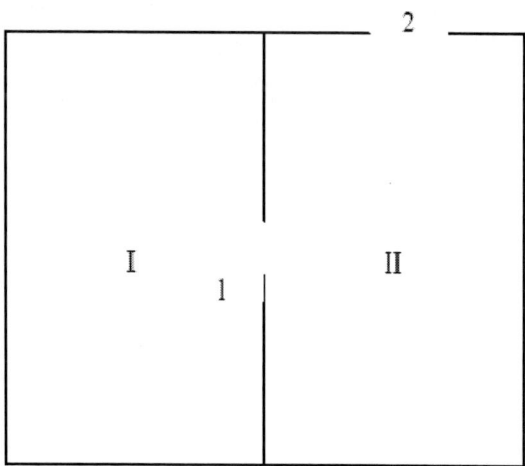

Figure 18. Schematic of a double effusion chamber. Chamber: (*I*) evaporation and (*II*) effusion; (*1*) connecting channel and (*2*) effusion orifice.

Under our experimental conditions, changes in the mixture composition during measurements were negligible. Thus, the whole body of the data obtained can be related to the state of the system, which is characterized by the starting composition. A solution to be analyzed was placed in chamber *I* (evaporation chamber). After the solution was heated to a temperature *T*, vapors of its components passed through a connecting channel into chamber *II* (effusion chamber), where the effusion occurred. With this design of the effusion unit, the pressure in the effusion chamber $\sum_{i=1}^{n} p_{2i}$ at a given temperature *T* will be determined by the vapor pressure in the evaporation chamber, $\sum_{i=1}^{n} p_{1i}$, and by the relative throughputs of the connecting channel ($S_l L_l$) and effusion orifice ($S_{ef} L_{ef}$). The flux ρ_1 arriving into the effusion chamber is given by

$$\rho_1 = S_l L_l \sum_{i=1}^{n} A_i \sum_{i=1}^{n} p_{1i}, \quad A_i = (2\pi M_i RT)^{-1/2}, \tag{11}$$

where ρ_1 is the flux of a mixture of *n* gases leaving the evaporation chamber in unit time (mol m^{-2} s^{-1}); S_l, cross-sectional area of the connecting channel (m^2); L_l, Clausing coefficient characterizing the resistance of the connecting channel to a vapor flow; M_i, molecular mass of *i*th component of the mixture (kg mol^{-1}); R = 8.314 J mol^{-1} K^{-1}, universal gas constant; *T*, absolute temperature (K); $\sum_{i=1}^{n} p_{1i}$, total pressure of a mixture of *n* gases in the evaporation chamber (Pa); and p_{1i}, partial pressure of *i*th component of the mixture in the evaporation chamber (Pa). The flux ρ_2 leaving the effusion chamber is given by

$$-\rho_2 = S_l L_l \sum_{i=1}^{n} A_i \sum_{i=1}^{n} p_{2i} + S_{ef} L_{ef} \sum_{i=1}^{n} A_i \sum_{i=1}^{n} p_{2i}, \tag{12}$$

where S_{ef} and L_{ef} are the cross-sectional area of the effusion orifice and the Clausing coefficient, respectively, and p_{2i} is the partial pressure of *i*th component of the mixture in the effusion chamber (Pa). In the steady-state mode, the fluxes are equal ($\rho_1 = \rho_2$) and we have

$$\sum_{i=1}^{n} p_{2i} = [S_l L_l / (S_l L_l + S_{ef} L_{ef})] \sum_{i=1}^{n} p_{1i} \qquad (13)$$

The experiment can be performed in two ways. In the first case, a water-DMSO solution with a known concentration of the components is placed in chamber *I*, and water, in chamber *II*. If chamber *II* contains pure water, analysis of Eqs. (11)-(13) shows that, in this case, the water vapor pressures in chambers *I* and *II* will be the same and equal to the saturated water vapor pressure at the temperature of the experiment. This situation will be preserved while pure water is in chamber *II*. In this case, DMSO will evaporate from chamber *I*, and water will only evaporate from chamber *II* and will act as a seal that prevents evaporation from chamber *I*. In the second variant, pure DMSO is in the effusion chamber, and a two component water-DMSO solution of known concentration, in the evaporation chamber. In this case, the role of a seal is played by DMSO, which prevents evaporation of DMSO from chamber *I*, so that only water evaporates from chamber *I*. Generally speaking, it is necessary, when recording the fluxes p_1 and p_2, to make a correction for the transfer of a component from a chamber with a lower partial pressure into that with a lower partial pressure, because the concentrations of the solutions vary in the closed isolated space in such a way that the vapor pressures over the solutions tend to equalize. However, it is shown below that this correction can be neglected under the conditions of a not-too-long effusion experiment. The dismountable design of the effusion unit makes it possible to record the loss of mass by a solution in each chamber during the experiment and to vary the throughput capacities of the connecting channel and effusion orifice to achieve the optimal conditions of the experiment. The loss of mass by the solution, temperature of the effusion unit, and evaporation time were used to calculate the vapor pressure in the effusion and evaporation chambers by Knudsen's equation. The vapor pressure of *i*th component in the effusion chamber, p_{2i} (Pa), is given by

$$p_{2i} = (\Delta m_i / \alpha \tau S_{ef} L_{ef})(2\pi M_i RT)^{1/2}, \qquad (14)$$

where Δm_i is the mass of substance effused from the effusion unit (kg); τ, evaporation time (s); and α, condensation coefficient.

Then, with account of Eq. (13), we can express the vapor pressure of *i*th component in the evaporation chamber as

$$p_{li} = (\Delta m_i/\alpha\tau S_{ef}L_{ef})[(S_lL_l + S_{ef}L_{ef})/S_lL_l](2\pi M_i RT)^{1/2} \qquad (15)$$

Thus, measuring the loss of mass by a solution in the evaporation chamber in two variants of the experiment, we can obtain the partial pressures of the solution components in relation to concentration. We used in the experiment an effusion orifice with a throughput $S_{ef}L_{ef}$ of $8.206 \cdot 10^{-10}$ m^2. The throughput of the connecting channel S_lL_l was varied as follows: $6.6 \cdot 10^{-11}$; $5.66 \cdot 10^{-10}$; $1.092 \cdot 10^{-9}$ m^2. The procedure described in detail in [33] was used to construct a calibration plot relating the water vapor pressure calculated by Knudsen's equation (15) to the true pressure. It was also found that the condensation coefficient α in Eq. (15) is 0.63 for pure DMSO. Further, we used this value to calculate partial pressures of DMSO in water-DMSO solutions. It was found that the partial pressures of water and DMSO obtained at 298 K using the procedure described above and the published data [23] are in very good agreement [34].

In addition, we determined the equilibrium partial pressures of water and DMSO over the cellulose-water-DMSO system. The procedure used to measure the water vapor pressure over cellulose samples was described in detail previously. With account of the specific features of analysis of a two-component water-DMSO solution, the method consists in the following. A preliminarily dried cellulose sample is wetted with a minor amount of the water-DMSO solution and an effusion experiment is carried out using a double effusion chamber. In the experiment, the partial pressures of water and DMSO vapors were determined at 298 K over a cellulose sample wetted with water-DMSO solutions of various concentrations. The content of a component in cellulose, a_i, at which the equilibrium vapor pressure p_i of the component starts to deviate from its saturated vapor pressure p_{i0} at the temperature of the experiment (298 K), i.e., the ratio p_i/p_{i0} starts to deviate from unity, corresponds to the maximum sorption of the component by cellulose at a given concentration of the solution [33]. In this way, we obtained individual isotherms of water and DMSO sorption by cellulose from a water-DMSO mixture at 298 K.

Figure 19 shows individual isotherms of water and DMSO sorption by cellulose from a water- DMSO mixture at 298 K, obtained in three different ways: (1) by determining the total absorption of the solution by cellulose (cellulose swelling) [35], (2) by calculated from an experimental isotherm of composition variation, and (3) by the method using double effusion chamber.

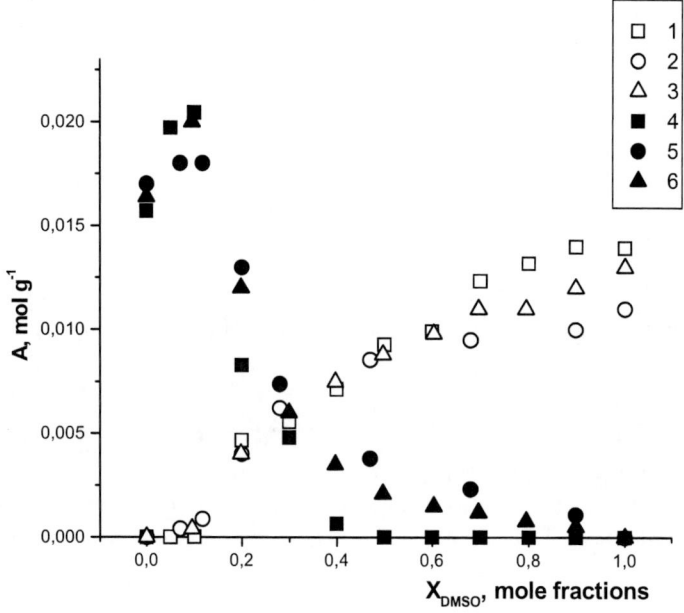

Figure 19. Individual isotherms of (*1-3*) DMSO and (*4 - 6*) water sorption A (mol g^{-1}) from a water-DMSO mixture on cellulose at 298 K. (*1, 4*) Obtained by determining the total absorption of the solution by cellulose; (*2, 5*) calculated from an experimental isotherm of composition variation; (*3, 6*) obtained using double effusion chamber.

The method for calculating individual sorption isotherms of components from the isotherm of variation of the mixture composition has been described in detail in the literature [36], whereas the method for determining the total absorption of a solution by cellulose requires a more detailed description. A sample of cellulose wetted with a water-DMSO solution was centrifuged at 8000 rpm for 15 min. These conditions make it possible to remove the solution mechanically bound to cellulose and obtain a cellulose sample with the content of components equal to the physical sorption on cellulose [35]. The content of DMSO in the cellulose sample was determined by extraction of DMSO into a fixed volume of water. The concentration of the resulting solution was found by refractometry. The water content of the sample was calculated from the total mass of the solution sorbed by cellulose (after the centrifugation) and the DMSO content of the sample. It can be seen from figure 19 that the dependence of water sorption by cellulose from a binary water-DMSO solution has an extremum: the maximum sorption of water is observed at DMSO concentrations X_{DMSO} = 0.05-0.1 mol fraction.

The obtained individual isotherms for water and DMSO adsorption on cellulose indicate that in the range of $0<X_{DMSO}<0.28$ the water sorption dominates. In the presence of small amounts of DMSO ($X_{DMSO}\sim0.1$) water sorption is higher than that for pure water. Increasing of X_{DMSO} in the range of 0.07–0.1 leads to breakage of water clusters and forming of hydrophobic DMSO dimers [27]. As it was mentioned above, the water evaporation enthalpy remains constant in this concentration range but the change of microscopic solvent structure at $X_{DMSO}\sim0.1$ probably leads to the increasing of water sorption. This finding is in good agreement with the NMR observation of the anomalous increase of motional freedom of water molecules at $X_{DMSO} \geq 0.1$ [29]. The behaviour of individual adsorption isotherms of water and DMSO changes dramatically at $X_{DMSO}\sim0.28$. In the range of $0< X_{DMSO} <0.28$ water sorption dominates, whereas in the region of $X_{DMSO} \geq 0.28$ DMSO sorption dominates and it remains nearly constant with increasing of DMSO concentration. It is worth noting that $X_{DMSO}\sim0.28$ is the concentration boundary mentioned above and observed in various experimental studies. In the water-rich region ($X_{DMSO}<0.28$) DMSO dimers exhibit hydrophobic properties, whereas in the region of $X_{DMSO} >0.28$ DMSO clusters do not have any hydrophobic properties. In the region of $X_{DMSO} <0.28$ the hydrogen bonding between the cellulose hydroxyls and DMSO oxygen atom is an unfeasible process, therefore the value of DMSO sorption on cellulose is expected to be insignificant and water sorption dominates. In the DMSO-rich region $X_{DMSO} >0.56$, DMSO sorption dominates, whereas water sorption is practically insignificant. The immersion heats of cellulose in water–DMSO mixtures and desorption enthalpies of water and DMSO for the water–DMSO–cellulose system are shown in figures 20 and 21, respectively.

The curves have the same character. The enthalpy curves demonstrate the minimum at $X_{DMSO}=0.6$ for water and at $X_{DMSO} =0.4$ for DMSO. We can assume that decreasing of desorption enthalpy of water in the range of $X_{DMSO} <0.6$ related to the dehydration process because the adsorption from the solution has a displacing nature. Increasing of desorption enthalpy of water in the range of $X_{DMSO}>0.6$ can be explained by disappearing of strong interacting single water molecules with DMSO clusters. Increasing of DMSO desorption enthalpy in the range of $X_{DMSO} >0.4$ (see figure 21) is presumably related to the specific interactions between cellulose and DMSO. Such interactions can be described in terms of hydrogen bonding between the cellulose hydroxyls and DMSO oxygen atom. In the range of $X_{DMSO}<0.28$, intermolecular interactions within DMSO molecules are much stronger than for the case of the DMSO-rich region (see figure 15).

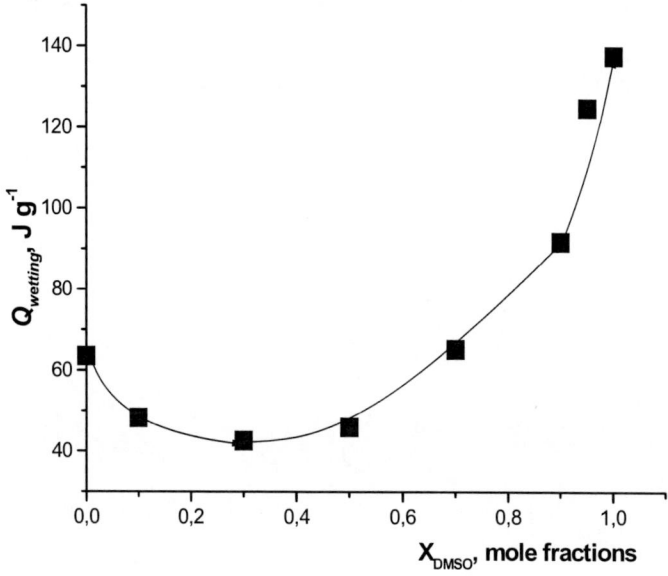

Figure 20. The immersion heats of cellulose in water–DMSO mixtures.

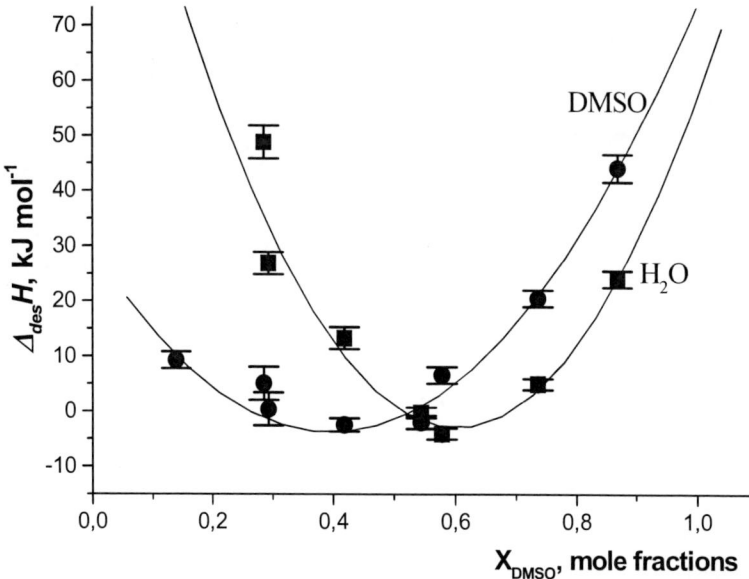

Figure 21. The desorption enthalpies of water and DMSO from the aqueous DMSO–cellulose system.

Therefore, at X_{DMSO}<0.28 specific interactions between DMSO and cellulose are less probable. This conclusion is in a good agreement with the individual adsorption isotherms of water and DMSO from the water–DMSO mixtures on cellulose. The complex behaviour of the individual adsorption isotherms of water and DMSO, the desorption enthalpies of water and DMSO, and the immersion heats of cellulose demonstrate that the solvation process of cellulose is nonlinearly related to the mixing ratios of DMSO–water mixtures. Hence, the microscopic solvent structure in the water–DMSO mixtures influences considerably the energetic characteristics of interactions between water, DMSO and cellulose.

Chapter 3

ADSORPTION OF AQUEOUS-ORGANIC MIXTURE COMPONENTS ON CELLULOSE

Most processes for the isolation of cellulose from plant materials and physicochemical cellulose processing are based on the interaction of cellulose with low molecular weight liquids. In spite of a large number of works concerned with the sorption properties of cellulose, the question of the mechanism of its interaction with low molecular weight liquids remains open. To gain insight into this problem, we must determine the physicochemical characteristics of the interaction of the polymer with aqueous–organic mixture components and their relation to the structure of solutions.

DMSO and AN are aprotic bipolar solvents infinitely soluble in water. Their mixtures with water are, however, a good example of two nonideal binary mixtures with opposite properties. Various physicochemical properties of these binary mixtures are shown in figure 22. The form of the mixture composition dependences of various physicochemical properties is evidence of strong deviations from ideality caused by various intermolecular interactions. The deviation from a thermodynamically ideal binary mixture is caused by the microscopic structure of solutions, because water and organic solvent molecules mix with each other inhomogeneously at the level of clusters [27, 28]. The enthalpies of mixing show that the excess thermodynamic properties of these two mixtures are opposite, they are endothermic for AN–water and exothermic for DMSO–water mixtures [37]. Exothermic and endothermic mixing are enthalpy and entropy controlled processes, respectively, and interactions of water with DMSO and AN should therefore be substantially different.

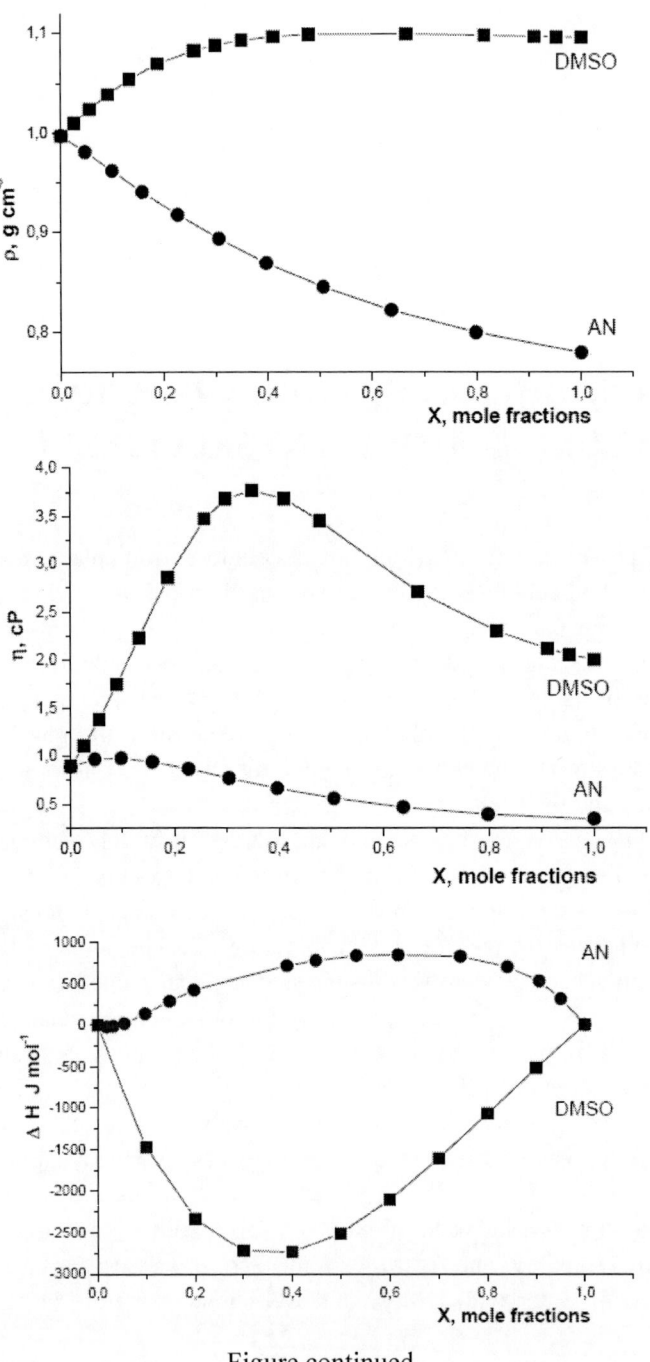

Figure continued

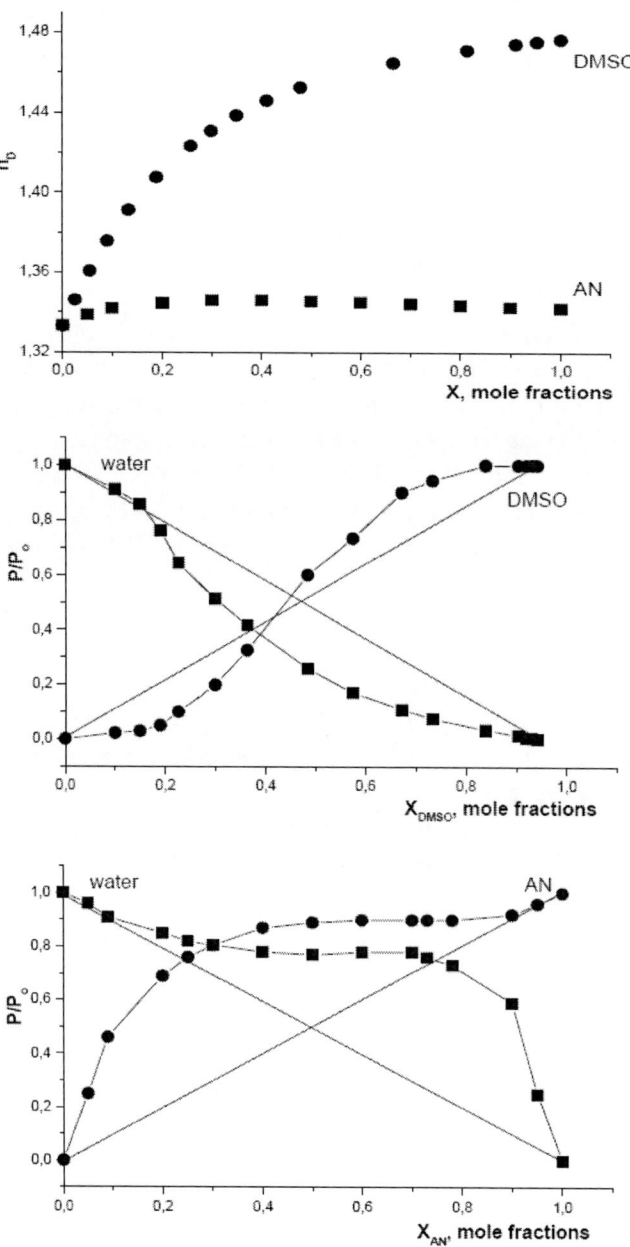

Figure 22. Physicochemical properties of water–DMSO and water–AN binary mixtures: density [24, 38], viscosity [24, 38], excess enthalpies of mixing [37], refractive index [24], and partial pressures [23, 39].

Some information about intermolecular interactions in binary mixtures can be obtained from deviations of the partial pressures of mixture components over solutions from the Raoult law. If molecules in a mixture interact less strongly than molecules in the pure substances, the partial pressures of the components over solution exhibit positive deviations from the Raoult law. Figure 22 shows that the partial pressures of both water and AN exhibit positive deviations from the Raoult law virtually over the whole range of AN–water mixture compositions. Over the concentration range $0.5 < X_{AN} < 0.8$ (mole fractions), the partial pressures of both components are virtually independent of concentration. This means that each molecule of those present in the mixture more effectively interacts with like molecules than with molecules of a different kind. This situation is possible if water molecules are bound into three-dimensional clusters by H-bonds, and these clusters are surrounded by AN molecules. AN molecules behave as a "solvent" and can be oriented toward water clusters by dipole–dipole interactions [39]. In the water–DMSO system, the partial pressure of one of the components (water) exhibits negative deviations from the Raoult law, whereas deviations of the partial pressure of DMSO change sign. The latter circumstance presupposes not only the presence of complex intermolecular interactions in the water–DMSO system but also a change in the mean size of associates as the mole fraction of DMSO increases. In [39, 27, 28, 40, 41], the composition of clusters formed in the water–DMSO and water–AN systems was studied using a special mass spectrometric technique for direct isolation of clusters from a mixture of solvents. According to [39, 27, 28, 40, 41], water clusters exist in the water– DMSO system only up to the concentration $X_{DMSO} < 0.07$, and they can interact with one or two DMSO molecules. The number of water clusters decreases sharply at $0.07 < X_{DMS} < 0.1$ and then slowly at 0.1–0.28 mole fractions; simultaneously, the number of DMSO clusters increases [27]. The network of water H-bonds breaks down. DMSO exists in the form of dimers or "small" clusters, in which DMSO molecules are held together by S=O dipoles that have antiparallel orientations [27, 26]. Hydrophobic CH_3-groups are situated on the outside, and DMSO clusters therefore have hydrophobic properties. Hydrophobic behavior of DMSO over this concentration range was substantiated by X-ray structure data [26]. Over the concentration range $0.28 < X_{DMSO} < 0.6$, the size of DMSO clusters increases and, at $X_{DMSO} > 0.6$, DMSO molecules form clusters of the same structure as in pure DMSO. The addition of AN to water in concentrations $0 < X_{AN} < 0.2$ decreases the number of free water molecules substantially. In other words, the fraction of water clusters increases as the concentration of AN grows. According to [39, 40], the addition of AN to water stabilizes the network of water H-bonds and increases the lifetime of large clusters. At $X_{AN} > 0.2$, a large number of AN molecules stabilize water clusters and suppress the thermal motion of water molecules because of the formation of a strong network of bonds. Water clusters exist up to the concentration X_{AN} ~0.8; at higher AN concentrations, the network of H-bonds breaks down. At $X_{AN} > 0.8$, AN clusters containing AN molecules only are formed [41]. To summarize, it was shown in several works that the modes of interactions between water and organic component molecules were substantially different in character for water–DMSO and water–AN systems.

The isotherms of composition changes and the calculated isotherms of component sorption from water–DMSO and water–AN solutions by cellulose are shown in figures 23 and 24.

The isotherms of component sorption obtained show that the sorption characteristics of water–DMSO and water–AN mixture components are substantially different. Note that water is differently sorbed on cellulose in the presence of different organic components. For the cellulose–water–DMSO system at concentrations $0 < X_{DMSO} < 0.28$, the sorption of water on cellulose prevails; in the presence of small amounts of DMSO (X_{DMSO} ~0.1), water is adsorbed in larger amounts than from pure water. For the cellulose– water–AN system, water sorption sharply decreases already in the presence of small amounts of AN; at concentrations higher than X_{AN} ~ 0.2, the sorption value remains virtually constant.

It can be suggested that the difference in the character of the isotherms of sorption of water on cellulose from two aqueous–organic mixtures is determined by the character of cluster formation. Water [42], AN [40], and DMSO [28] easily form clusters in the pure state. The addition of an organic solvent to water, however, destroys AN clusters in water–AN mixtures and water clusters in water–DMSO mixtures. In water–AN mixtures, water clusters exist up to the concentration X_{AN} ~ 0.8 because of their stabilization by AN molecules [39]. Conversely, in water–DMSO mixtures, water clusters decompose already at the concentration X_{DMSO} ~ 0.07–0.1. The destruction of water clusters at concentrations $X_{DMSO} > 0.07$–0.1 was related to the formation of DMSO dimers with hydrophobic properties [28]. Smaller sorption of water on cellulose from water–AN solutions compared with water–DMSO mixtures is likely related to the stabilization of water clusters by AN molecules. AN molecules very weakly interact with cellulose; they surround water clusters and prevent them from interaction with cellulose. The sorption of water from water–DMSO mixtures is higher, and the presence of a maximum at xDMSO ~ 0.1 is related to the existence of DMSO in the form of hydrophobic dimers.

Note also that the character of sorption on cellulose of DMSO and AN from their mixtures with water is different.

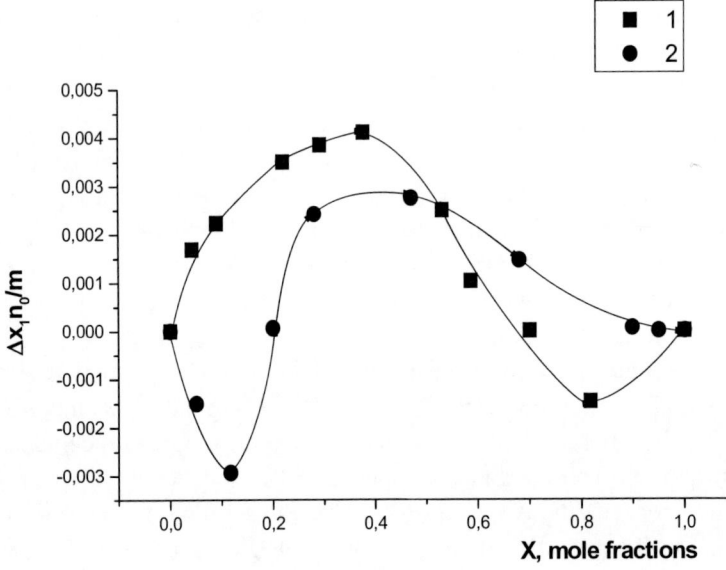

Figure 23. Isotherms of changes in the composition of (*1*) water–AN and (*2*) water–DMSO solutions in the presence of cellulose.

This is especially clearly seen for the initial isotherm portions (figure 24). Up to the concentration $X_{DMSO} \sim 0.1$, sorption of DMSO is virtually zero. In this concentration region, DMSO forms hydrophobic dimers, which are not likely to be adsorbed on cellulose. At higher concentrations, the sorption of DMSO gradually increases and reaches a constant value of 0.01 mol/g at the concentration $X_{DMSO} \sim 0.6$. The sorption of DMSO increases as the size of DMSO clusters gradually grows and their hydrophobic properties weaken. The size of DMSO clusters reaches its constant value corresponding to the structure of pure DMSO also at $X_{DMSO} \sim 0.6$. The isotherm of sorption of AN has the shape of a Langmuir isotherm and reaches saturation, 0.0028 mol/g, already at the concentration $X_{AN} \sim 0.1$. AN molecules do not form clusters in solutions enriched in water, and AN behaves as a "solvent," which surrounds and stabilizes water molecules. The difference in interactions of water–DMSO and water–AN mixtures with cellulose influences such characteristics as the heats of wetting and the total absorption of solution by cellulose (figures 25, 26).

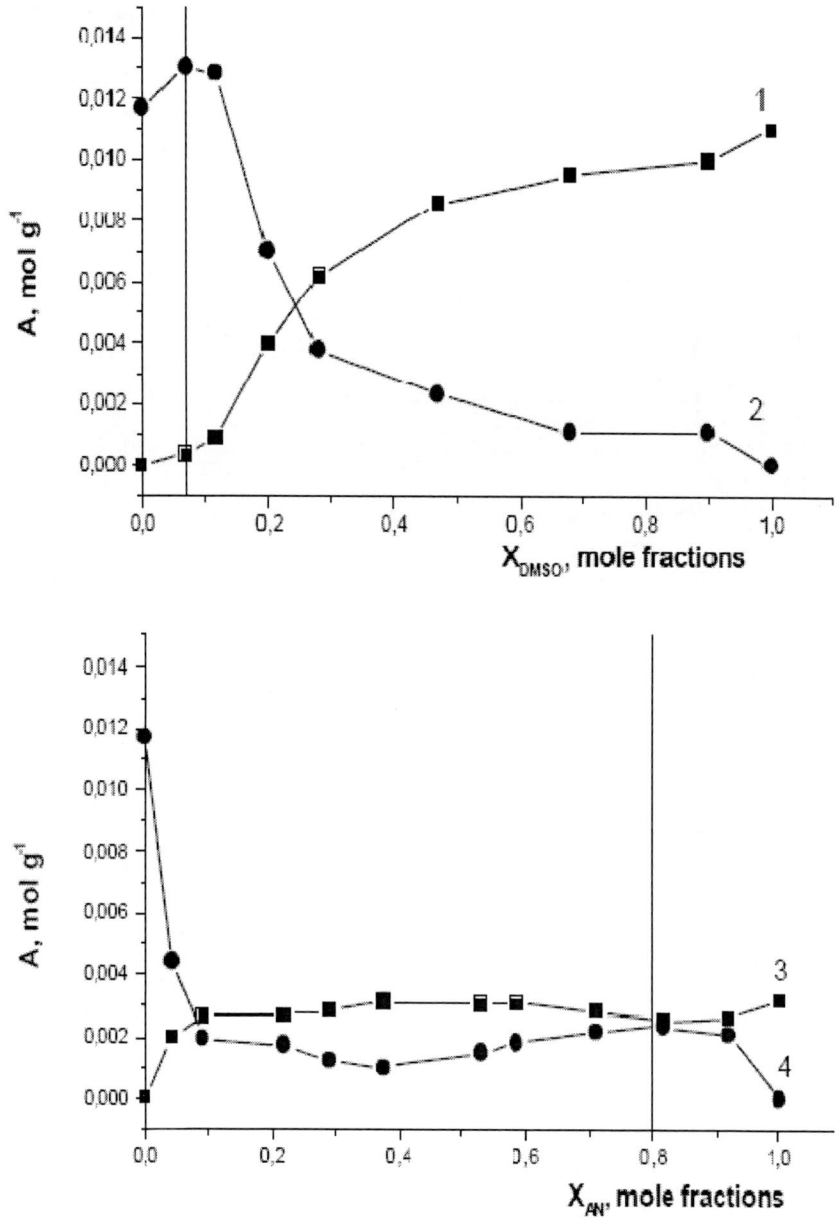

Figure 24. Isotherms of adsorption of the components of water–DMSO ((*1*) DMSO and (*2*) water) and water–AN ((*3*) AN and (*4*) water) solutions on cellulose. The vertical line is the boundary of the existence of water clusters.

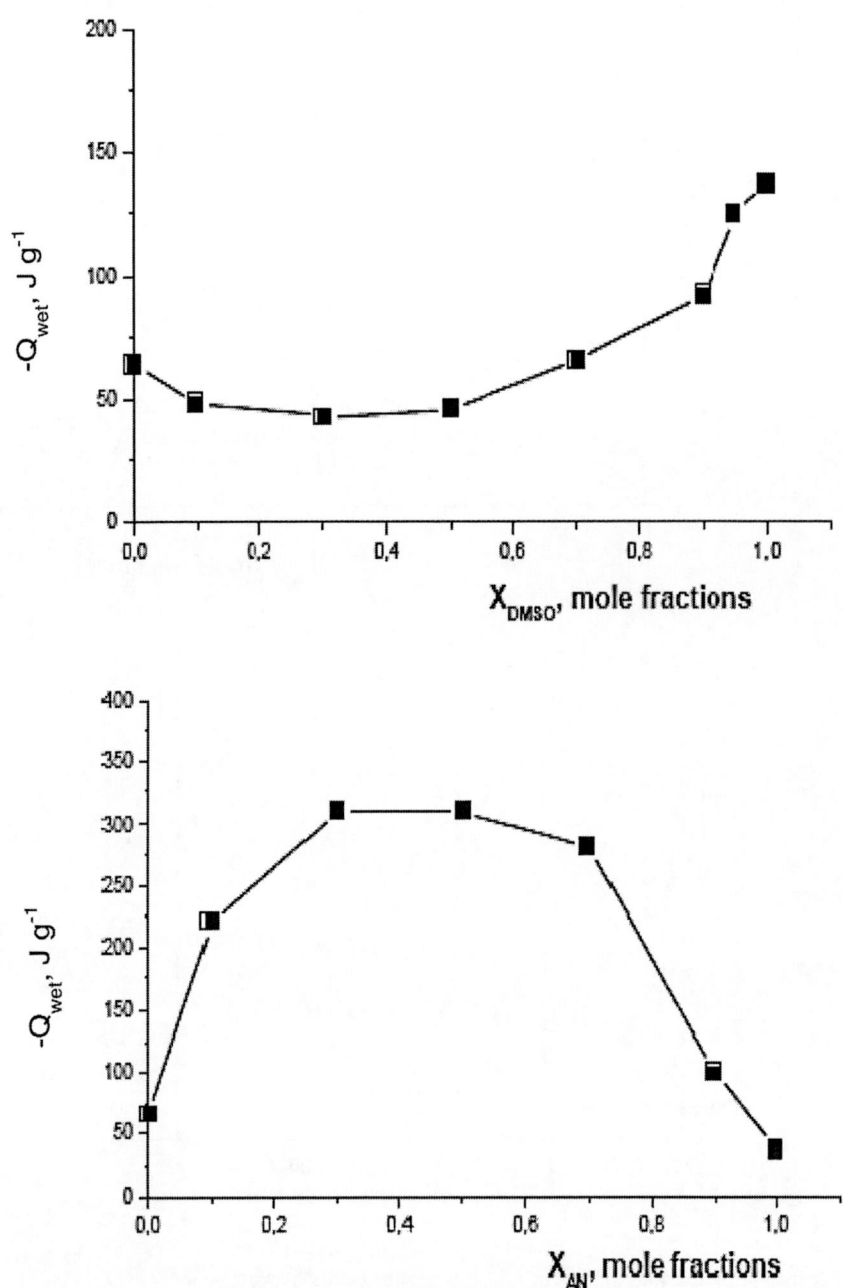

Figure 25. Heats of wetting of cellulose with water–DMSO and water–AN mixtures.

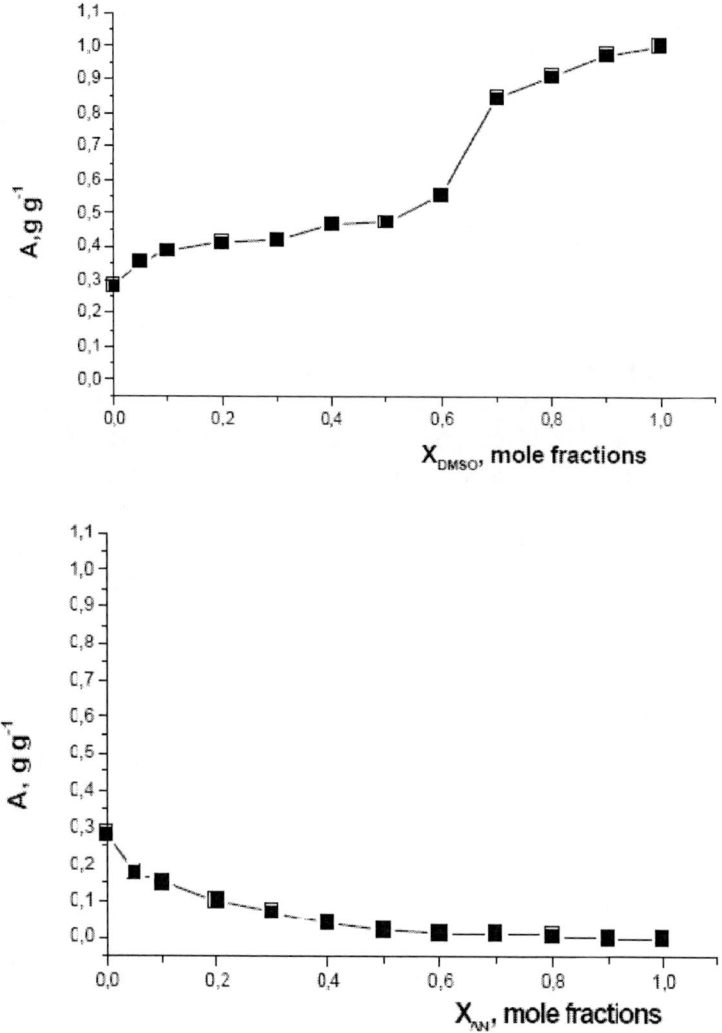

Figure 26. Total absorption of water–DMSO and water–AN solutions by cellulose.

The heats of wetting of cellulose by water–DMSO and water–AN mixtures were determined on a liquid isoperibolic calorimeter by breaking a glass ampule with a polymer sample in a reaction beaker filled with mixed solvents. The determination of the total absorption of solution by cellulose (the degree of cellulose swelling) was performed as follow [35]. A wet cellulose sample was centrifuged at a 8000 rpm rate for 15 min. This allowed us to obtain cellulose with the moisture content equal to that resulting from the sorption of water vapor from the gas phase. The same centrifuging conditions were used with cellulose samples wetted with binary aqueous–organic solvents.

The heat of wetting is an integral value including the heat effects of all processes that occur in the interaction between polymers and solutions. This value can include not only the heat effect of sorption (exothermic effect), but also the heat effect of cellulose swelling (endothermic effect) related to changes in intermolecular interactions between macromolecules and supramolecular formations in the polymer and the heat effect caused by changes in the structure of the solvent in the surface layer of a finite thickness. According to figure 26, the weight of cellulose increases in the water–DMSO–cellulose system as the mole fraction of DMSO grows because of the absorption of the liquid (cellulose swelling occurs). In the water–AN–cellulose system, the weight of wetted cellulose decreases as the mole fraction of AN increases; that is, swelling of cellulose in water–AN mixtures is smaller than in pure water, and swelling in pure AN is virtually zero. It follows that unexpectedly larger heat effects of cellulose wetting with water–AN mixtures can be explained by the absence of the endothermic contribution to the total heat effect of cellulose swelling. In addition, it can be suggested that the sorption of water from water–AN mixtures occurs with the destruction of shells of AN molecules, which stabilize water clusters in solution. This can make an additional exothermic contribution to the heat effect of cellulose wetting with water–AN mixtures (because the heat effect of mixing of water with AN is endothermic). On the other hand, the total heat effect of wetting cellulose with water–DMSO mixtures can include an endothermic component related to changes in intermolecular interactions between macromolecules in cellulose (swelling). Probably for this reason, the heat effects of wetting in the water–DMSO–cellulose system are on the whole lower. An increase in the heat effect of wetting of cellulose at mixture compositions $X_{DMSO} > 0.6$ can be explained by specific interactions between cellulose and DMSO, that is, the formation of H-bonds between cellulose hydroxyls and DMSO oxygen atoms. At lower concentrations at which DMSO exists in the form of hydrophobic dimers or small clusters with CH_3-groups on the outside, specific interactions of this kind are impossible [43].

It follows that intermolecular interactions in the water–DMSO and water–AN systems determine the sorption of the components of these binary aqueous–organic mixtures on cellulose. The destruction of water clusters and formation of DMSO clusters, whose size and concentration increase, occurs in the cellulose–water–DMSO system as the content of DMSO grows. As a result

Chapter 4

ADSORPTION OF PHENOL AND TOLUENE FROM THE GAS PHASE AND AQUEOUS SOLUTIONS ON CELLULOSE

Sorption from solutions on polymeric materials forms the basis of many physicochemical processes. In particular, these are processes associated with the vital activity of a living body, which depends on the capability to sorb through respiratory and digestive organs and through skin various substances that stimulate or prevent normal functioning of the body. Industrial processes also involve accumulation of certain compounds and utilization of by-products. Polymeric sorbents exhibit certain specific features governed by their properties and structure. On the molecular level, these materials are mixtures of macromolecules of different lengths and irregular structure. The supramolecular structure of the polymers is characterized by the existence of areas significantly differing in the extent of macromolecular ordering. To a first approximation, sorbing amorphous regions and sorption-inactive crystallites can be distinguished [44]. Despite a large number of papers dealing with sorption properties of polymers, the mechanism of the interaction of polymers with mixtures of low molecular weight liquids (solutions or emulsions) is still poorly understood. In this section we consider sorption of two aromatic compounds differing in the nature of substituents (hydrophilic in phenol, hydrophobic in toluene) from the gas phase and aqueous solutions on various celluloses differing in the ratio of amorphous and crystalline phases.

To obtain adsorption isotherms, we used the isothermal saturation method. The concentration of organic substances in water before (c_0) and after (c_e)

sorption was determined by UV spectrophotometry. The adsorption value c_p was calculated by the equation

$$c_p = (c_0 - c_e)V/m, \qquad (16)$$

where c_0 and c_e are the initial and equilibrium sorbate concentrations, respectively (M); V, solution volume (l); and m, adsorbent weight (g).

To determine the water retention, a wet cellulose sample was centrifuged at 8000 rpm for 15 min [35]. Such treatment yields a cellulose sample with the water content virtually equal to that obtained by sorption of water from the gas phase. The water retaining power (%) was determined by the formula

$$W_{sp} = (m_2/m_1) \times 100\%, \qquad (17)$$

where m_1 is the dry cellulose weight (g) and m_2, cellulose weight after centrifugation (g).

Adsorption of toluene and phenol from the gas phase was monitored gravimetrically. A cellulose sample was placed in a desiccator with the atmosphere saturated with toluene or phenol vapor. The equilibrium was attained within 17–35 days.

The physicochemical properties of the celluloses used as adsorbents and the maximal values of phenol and toluene adsorption form the gas phase are given in Table 1.

The cellulose samples in hand differ in the raw material, preparation procedure, and chemical composition. Flax cellulose was prepared from intermediate flax by the alkali–peroxide procedure. Sulfite wood pulp [GOST (State Standard) 3914–89] was the starting material for preparing sample nos. 3 and 4 by chemical and mechanical processing. Cotton microcrystalline cellulose met TU (Technical Specification) 64-11-129–92. As seen from Table 1, adsorption of toluene from the gas phase is virtually independent of the kind of cellulose. Adsorption of phenol is 3–4 times higher than that of toluene and only slightly increases with an increase in the relative amount of the amorphous phase in cellulose. The pattern of the sorption of organic compounds from water on cellulose is complex. This is due to the fact that both components of the binary solution water–organic component interact with cellulose. Sorption occurs by the displacement mechanism on the surface of pores of cellulose swollen in water. The toluene and phenol adsorption isotherms that we obtained (figure 27) were described by the equation of the theory of volume filling of micropores [45–47].

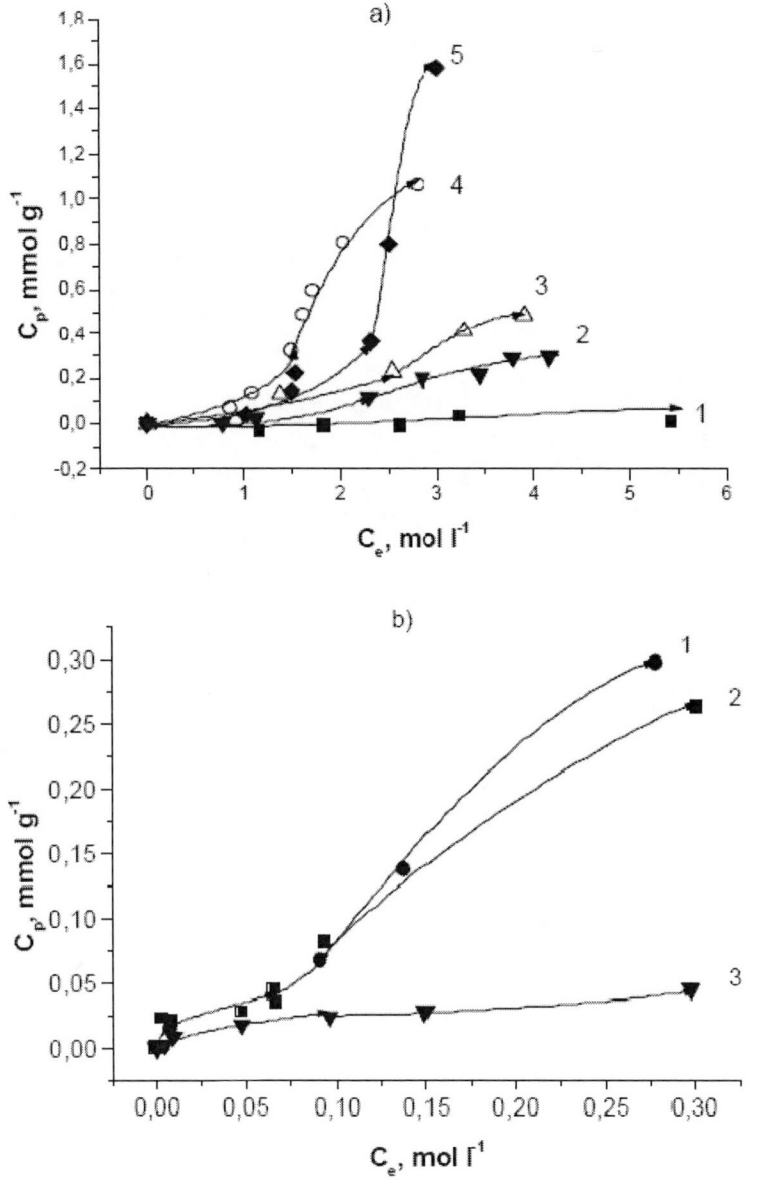

Figure 27. Isotherms of adsorption (c_p, mmol g–1) of (a) toluene and (b) phenol from water on celluloses. (c_e, mol l–1) Equilibrium concentration. Figures at curves are sample nos. in Table 1.

Table 2 presents the parameters of the Dubinin–Radushkevich equation for sorption from solutions:

$$\ln c_p = \ln c_\infty - (RT/E)^n [\ln(c_s/c_{sol})]^n, \quad (18)$$

where c_p and c_{sol} are the sorbate concentrations in the polymer (mol g–1) and solution (M), respectively; c_∞, limiting adsorption (mol g–1); c_s, limiting solubility of the sorbate in the given solvent (M); E, characteristic energy of adsorption; and n, coefficient showing by what factor the adsorption potential corresponding to the mean volume of sorbent pores is higher than the adsorption potential on the sorbent surface.

The correlation coefficient k of the dependence $\ln c_p - [\ln (c_s/c_{sol})]^n$ shows that Eq. (18) reasonably well approximates the experimental data. As can be seen (figures 27,28), phenol and toluene show essentially different dependences of the adsorption on the kind of cellulose. Celluloses that well adsorb toluene poorly adsorb phenol, and vice versa. Therefore, it can be assumed that phenol and toluene are adsorbed on different sites of cellulose. Since the degrees of crystallinity were determined not for all the celluloses studied, the fraction of the amorphous phase was characterized by the water retention [35]. This integral quantity characterizing the water content of cellulose depends on the content of noncellulose components, degree of degradation of polymer macromolecules, and kind of supramolecular structure.

To a greater extent, the water retention is proportional to the relative amount of the amorphous part, because the sorption of water and its interaction with hydroxy groups of cellulose occur specifically in the amorphous part. The dependences of the limiting sorption of phenol and toluene from water on the water retention of celluloses (figure 28) show that the adsorption of phenol increases, and that of toluene decreases, with an increase in the relative amount of the amorphous part. In adsorption of toluene from water (figure 28a), the adsorption value can be both higher (cellulose nos. 4, 5) and lower (cellulose no. 1), compared to the adsorption from the gas phase. Hence, the presence of water can both promote adsorption of toluene on cellulose (owing to diffusion) and prevent it.

Table 1. Physicochemical properties of cellulose samples studied

	Sample	Water retention, %	Moisture content of air-dry sample, %	Degree of crystallinity, %	Lignin content, %	Degree of polymerization	Adsorption from gas phase, mM	
							toluene	phenol
1	Flax cellulose, fiber	149.5	5.79	62	4.5	3000	0.414	1.548
2	Sulfite cellulose, fiber	145.4	5.88	67	0.1	800	0.389	1.186
3	Sulfite cellulose, powder, chemical disintegration	140.3	4.5	-	0.1	500	0.392	1.148
4	Sulfite cellulose, powder, mechanical disintegration	130.5	4.28	-	0.1	700	0.415	1.076
5	Microcrystalline cellulose (MCC)	129.6	4.59	73	0	250	0.403	1.100

Table 2. Parameters of the equation of the theory of volume filling of micropores for adsorption of organic compounds from aqueous solution at 298 K

S n	Sample	C_s, mmol l^{-1}	C_α, mmol g^{-1}	n	E, kJ mol^{-1}	k
Adsorption of toluene						
1	Flax cellulose, fiber	5.42	0.0098	2	1.216	0.999
2	Sulfite cellulose, fiber	5.42	0.300	2	2.399	0.994
3	Sulfite cellulose, powder, chemical disintegration	5.42	0.600	2	1.924	0.999
4	Sulfite cellulose, powder, mechanical disintegration	5.42	1.959	2	2.443	0.993
5	Microcrystalline cellulose (MCC)	5.42	2.534	2	1.983	0.984
Adsorption of phenol						
1	Flax cellulose, fiber	924.45	3.251	2	12.257	0.972
2	Sulfite cellulose, fiber	924.45	1.301	2	13.368	0.995
3	Sulfite cellulose, powder, chemical disintegration	924.45	0.200	2	15.799	0.993
4	Sulfite cellulose, powder, mechanical disintegration	924.45	0	-	-	-
5	Microcrystalline cellulose (MCC)	924.45	0	-	-	-

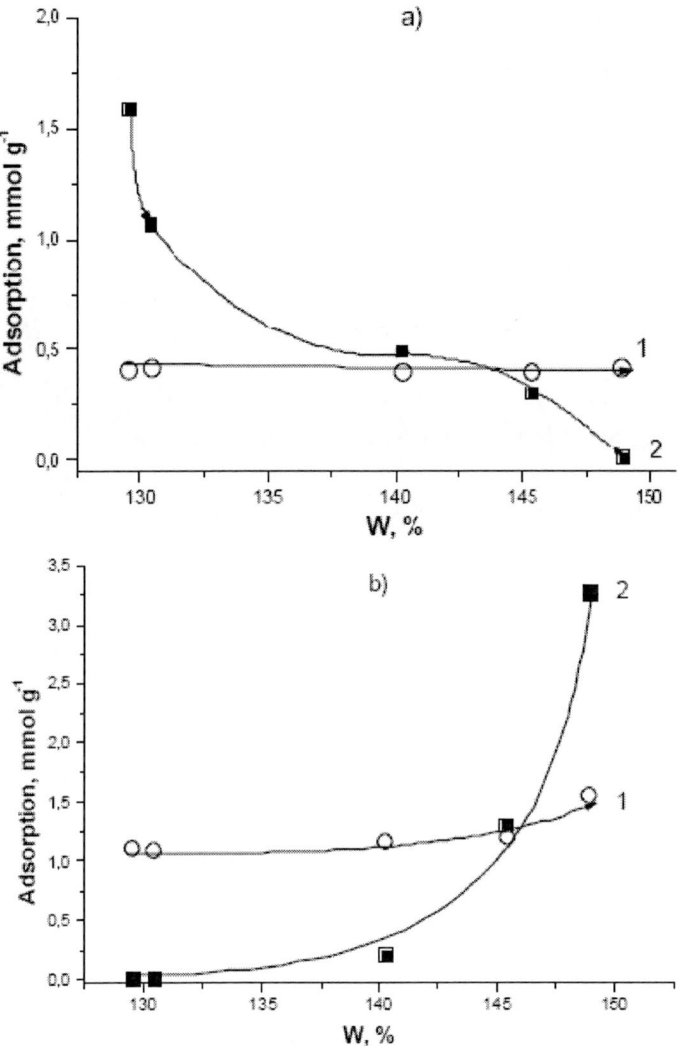

Figure 28. (*1*) Adsorption c_p (mmol g–1) of (a) toluene and (b) phenols from the gas phase and (*2*) c_∞___ (mmol g–1) limiting adsorption from water as functions of the water retention W (%) of cellulose nos. 1–5 (Table 1).

An increase in toluene adsorption in the presence of water is observed on cellulose samples (nos. 4, 5) with the maximally destroyed initial supramolecular structure of the fiber, high degree of crystallinity, and well pronounced surface of the crystallites. Presumably, displacement of water molecules with toluene molecules from the crystallite surface is more favorable energetically. A decrease in toluene adsorption in the presence of water is observed on the cellulose sample (no. 1) with the preserved fiber structure, in which the relative amount of the amorphous phase is larger than in microcrystalline cellulose. There is no interface between the amorphous and crystalline regions. Therefore, even in the presence of water the accessibility of the crystalline regions to toluene does not increase, and in the case of sample no. 1 it even decreases. Phenol is not noticeably adsorbed from aqueous solutions on cellulose samples with a high degree of crystallinity (sample nos. 4, 5). For sample nos. 1–3, with an increase in the relative amount of the amorphous phase, the phenol adsorption increases. Hence, it can be assumed that the phenol adsorption occurs in the amorphous part of cellulose. The characteristic energy of adsorption, calculated by Eq. (18), is different for adsorption of toluene and phenol. For the sorption of phenol, $E = 12$–16 kJ mol–1, which corresponds to the adsorption in micropores. For the sorption of toluene, E is lower by an order of magnitude, 1–2 kJ mol–1, which is more typical of surface sorption.

calorimeter. Based on the obtained data on phenol solubility and heats of phenol dissolution thermodynamic characteristics of phenol dissolution were determined (figure 30); here a pure substance having concentration of 1 mole fraction was used as a standard:

$$\Delta G^o_{sol} = -RT \lg K, \quad K = \frac{X_s}{X_o} \tag{19}$$

$$T\Delta S^o_{sol} = \frac{\Delta H^o_{sol} - \Delta G^o_{sol}}{T} \tag{20}$$

where X_s - solubility of phenol, mole fraction, $X_0=1$ – concentration of pure solute, mole fraction.

Phenol solubility in water-DMSO mixture increases sharply when DMSO concentration rises from 0.1 to 0.2 mole fraction (phenol solubility at $X_{DMSO}=0.1$ amounts to 0.322 mole per 100 g of solvent and at $X_{DMSO}=0.2$ is equal to 14.062 mole per 100 g of solvent). In the water-AN system phenol solubility decreases at AN addition, reaching the minimum at concentration =0.05. Then phenol solubility increases and reaches the value of solubility in pure water at AN concentration of about 0.22 mole fractions. Increase of AN concentration to more than $x_{AN} = 0.4$ mole fractions leads to high increase of phenol solubility just as in the case of water-DMSO mixture (phenol solubility at $X_{AN}=0.4$ amounts to 0.252 mole per 100 g of solvent and at $X_{AN}=0.5$ is equal to 17.587 mole per 100g of solvent).

At least two regions can be distinguished in thermodynamic correlations of phenol-water-DMSO system. The concentration range $0 > x_{DMSO} > 0.1$ is characterized by increase of endothermic effect of dissolution heats and by increase of entropy. This shows that effects of collapse of the existing structure prevail. In the interval $0.1 > x_{DMSO} > 1.0$ we observe decrease of endothermic effect of dissolution heats with the shift to exothermic region; this points out to strong interactions in the system followed by increase of entropy. It is known [27,32] that at concentration $x_{DMSO} > 0.1$ hydrophobic dimers of DMSO begin to form, which at further increase of concentration of organic component transfer to bigger clusters with gradual increase of their size. It can

Chapter 5

SORPTION OF PHENOL ON CELLULOSE FROM BINARY AQUEOUS-ORGANIC MIXTUR

Study and prediction of the properties of systems consist and an aqueous–organic solvent is necessary for solving pro particular, to the development of sorbents based on compos synthetic polymers for water treatment to remove aromatic petroleum products [23,24,37,38]. To understand the nature of low molecular weight liquids with a polymer, it is necess physicochemical features of the interaction of a polymer an aqueous–organic solution and the correlation of thes solution structure. The solvation effect shows a nonlinea component ratio, owing to the fact that the solvation s molecule or around functional groups of a polymer diffe from the bulk of the binary solvent. Deviations from ideal binary system arise when molecules of water an mixed nonhomogeneously on the level of clusters. He binary systems the preferential solvation occurs more cluster formation in binary solvents determines the so preferential solvation may be a key moment in sorption

This section provides a study of phenol dissolut organic mixtures water- DMSO and water- AN.

Phenol solubility at 298 K in the water-DMSO an measured by the method of isothermal saturation (fig dissolution in the above mentioned mixtures were

be supposed that from the moment of formation of DMSO clusters phenol-DMSO interactions begin to prevail.

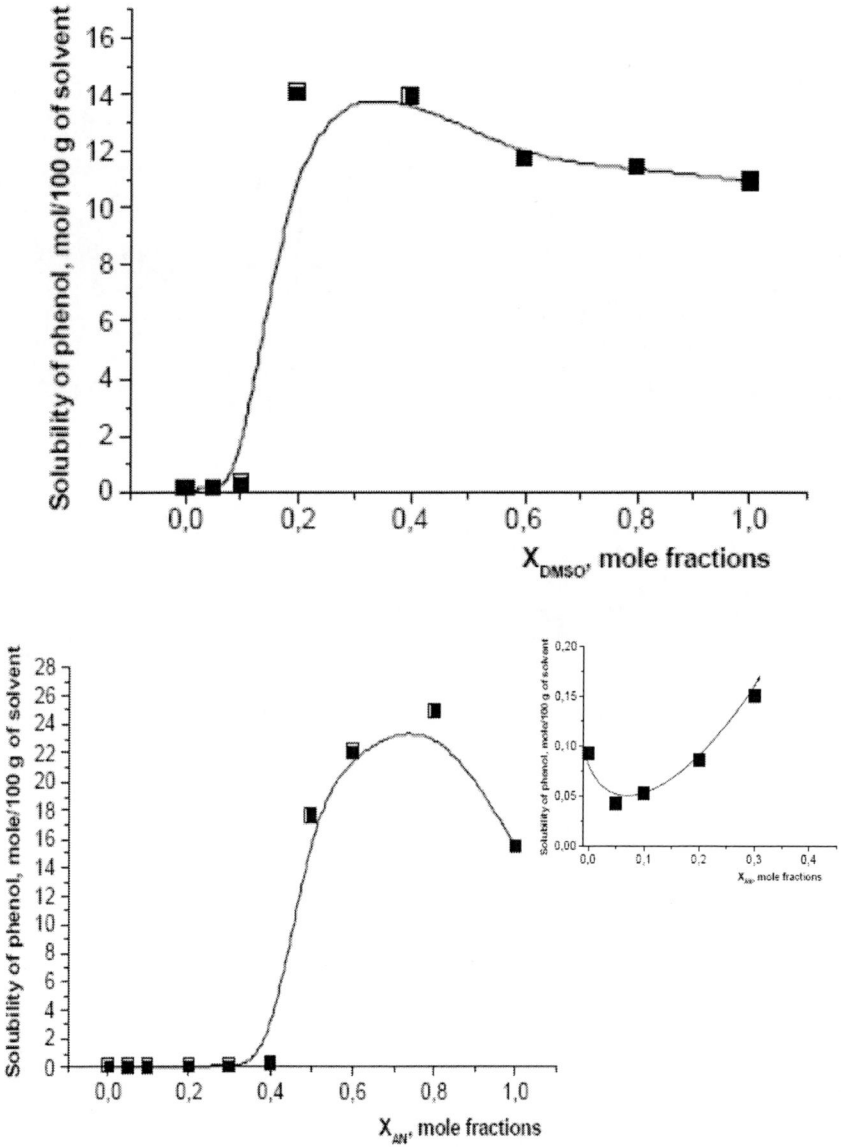

Figure 29. Solubility of phenol in mixtures water-DMSO and water-AN.

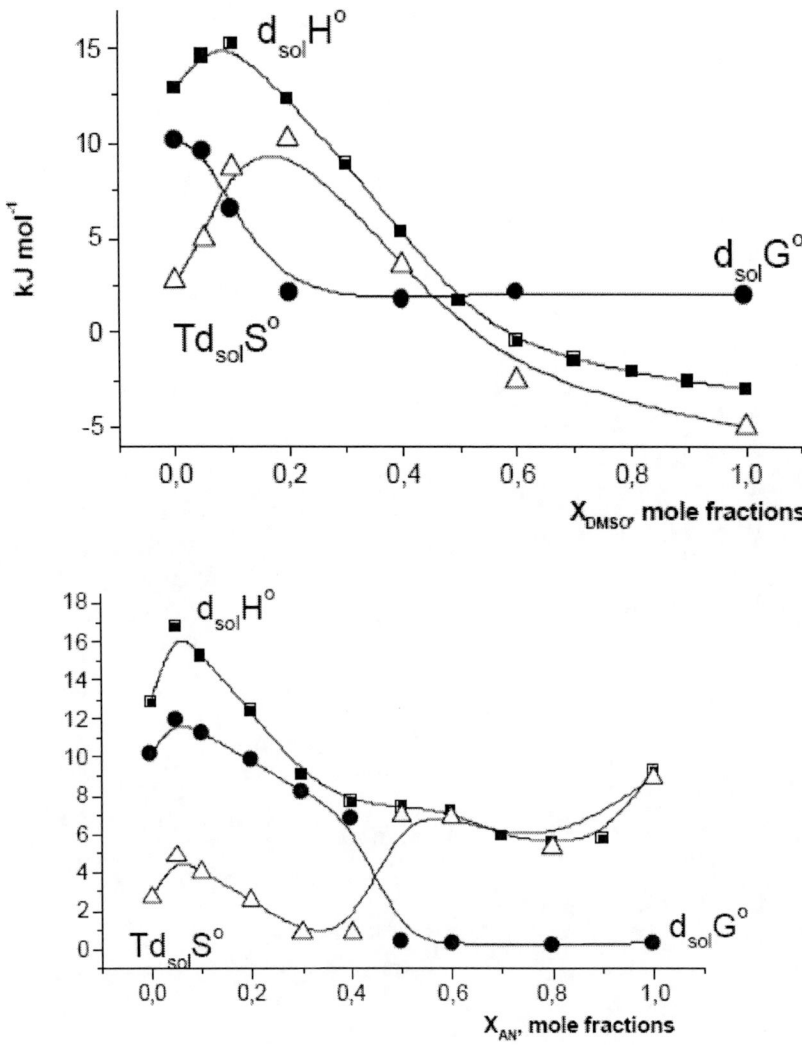

Figure 30. Dissolution thermodynamics of phenol in mixtures water-DMSO and water-AN.

Four regions can be distinguished in the phenol-water-AN system on the curves of thermodynamic relations. The region $0 > x_{AN} > 0.05$ (as for the phenol-water-DMSO system) is characterized by increase of endothermic effect of dissolution heats and by increase of entropy. In the range $0.05 > x_{AN} > 0.4$ we observe decrease of endothermic effect of phenol dissolution and decrease of entropy. Consequently, in this concentration range interaction processes followed by ordering of the system take place. It can be suggested (according to authors [26]) that this corresponds to the process of stabilization of water clusters by molecules of AN at x_{AN} >0.2, which limits thermal motion of water molecules due to strong network of hydrogen bonds. In this concentration range water-AN interaction prevails (in comparison to water-phenol and AN-phenol interactions). This leads to some decrease of phenol solubility in comparison with pure water.

In the interval 0.4 > x_{AN} > 0.8 increase of entropy takes place which is most likely dependant on appearance of free AN molecules not associated to the water clusters. This leads to sharp increase of phenol solubility. At concentration x_{AN} > 0.8 AN clusters consisting only of AN molecules are formed. Increase of entropy and endothermic effect of dissolution heats in this concentration range point out to processes of collapse of the existing structure during phenol dissolution.

It should be noted that in the range of low DMSO and AN concentrations (x ~0.05 mole fractions) when the structure of water is not collapsed yet under the influence of organic component, heats of phenol dissolution are practically similar in value for both water-organic systems. Besides that, appearance of the initial relational region is very similar. This allows suggesting that interaction mechanism of the solute with the components of the solvent is the same; this corresponds to solvation of phenol by water.

Thus, different character of cluster formation in water-DMSO and water-AN solutions gives rise the features of phenol solvation in these mixtures. Phenol dissolution in water-DMSO and water-AN mixtures depends on interactions with clusters existing in these binary solutions. At high water concentrations (x_{DMSO} <0.05, x_{AN} <0.4) phenol interacts with water clusters. The drastic increase of phenol solubility corresponding the increase of organic component content is due to interactions with formed DMSO and AN clusters.

Isotherms of phenol sorption from water and water-DMSO mixtures at three different temperatures are shown in figures 31 and32.

The data obtained show different types of sorption isotherms of phenol from water-DMSO x_{DMSO} = 0.05 and x_{DMSO} =0.3 mixtures. Type of isotherms from water-DMSO x_{DMSO} =0.05 mixture is analogous to sorption isotherms from water. In this case value for sorption is approximately twice as higher than the corresponding value for sorption from pure water. Isotherms of phenol sorption from water-DMSO x_{DMSO} =0.3 mixture represent curves with saturation. It can be suggested that different characters of isotherms is dependent on different character of phenol solvation in these concentration regions of the water-DMSO mixtures.

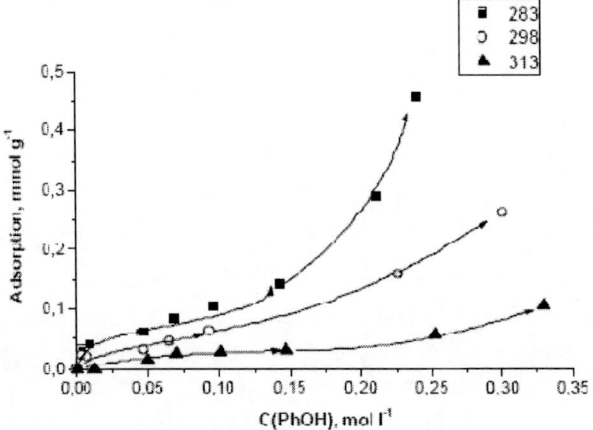

Figure 31. Isotherms of excess sorption of phenol on cellulose from water at various temperatures.

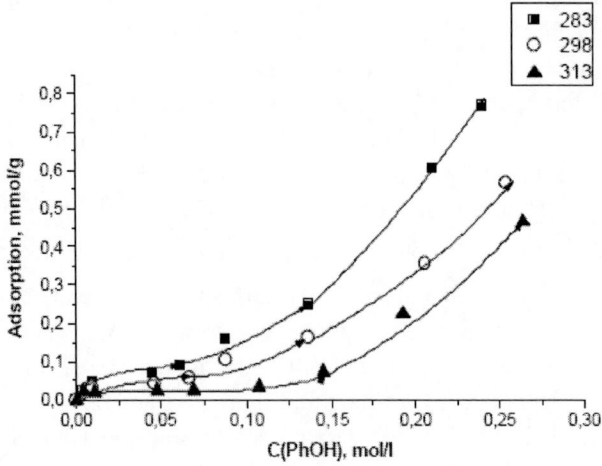

Figure 32. Isotherms of excess sorption of phenol on cellulose from water − DMSO (X_{DMSO} =0.05 and X_{DMSO} =0.3) mixtures.

Actually, as by authors shown [28], at these DMSO concentrations phenol is solvated differently: at concentration x_{DMSO} =0.05 it is predominantly solvated by water clusters, and at concentration x_{DMSO} =0.3 it is preferentially solvated by DMSO clusters, thus this defines the different characters of the isotherms.

The individual isotherms of both water and DMSO sorption on cellulose show that in the concentration range up to x_{DMSO} ~0.1 water sorption is higher than that from pure water (figure 24). In the same concentration range phenol sorption is higher than that from pure water. It can be suggested that in this case sorption of hydrated phenol takes place.

Process of phenol sorption from water-AN mixtures is more complicated (figure 33).

Phenol sorption from water-AN x_{AN} =0.05 mixture decreases with the increase of temperature, and in the case of water-AN x_{AN} =0.3 mixture it has anomalous temperature dependence. At temperature 283 K sorption of phenol is less than that at temperatures 298 and 313 K.

According to authors [39], in the concentration range $0 < x_{AN} < 0.2$ phenol interacts mainly with water clusters. At concentration higher than $x_{AN} > 0.2$ both water and phenol clusters are solvated by AN molecules so that interactions between water and phenol molecules are hindered. With the increase of temperature because of thermal motion the solvation shell of AN molecules collapses and interactions between water and phenol molecules become possible. This, in particular, defines increase of sorption of phenol on cellulose at temperature 298K. Further temperature increase up to 313 K leads to decrease of phenol sorption which is related to the total exothermic character of sorption. Combination of these factors: collapse of solvation shell of the solute (specific structural arrangement in the solution) and exothermic character of sorption leads to appearance of maximum point of sorption at definite temperature.

Consequently, the features of cluster forming in water-DMSO and water-AN mixtures and hence the features of phenol solvation determine phenol sorption on cellulose. The value of phenol sorption from water-DMSO mixture in the presence of small amounts of DMSO (x_{DMSO} ~0.05) is approximately two times higher than that from pure water similarly water sorption on cellulose from water-DMSO mixture. The value of phenol sorption from water-AN mixture decreases when AN concentration increases also similarly water sorption. In the concentration range $0 < x_{AN} < 0.2$ phenol interacts mainly with water clusters. In the range $x_{AN} > 0.2$ water clusters and phenol both are solvated with AN molecules that hinder from water-phenol interactions. Thus, phenol sorption on cellulose from binary water-DMSO and water-AN mixtures is determined first of all by phenol hydration. The solvation structure of phenol as a solute depends nonlinearly on the binary solvent composition due to preferential solvation caused by microscopic solvent structure. So, the preferential solvation is a key factor in sorption process on cellulose polymer.

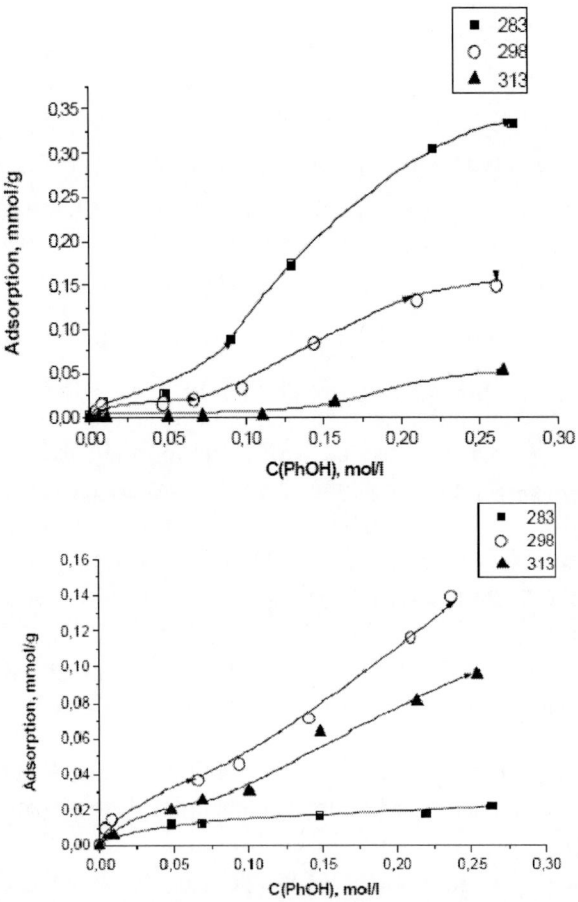

Figure 33. Isotherms of excess sorption of phenol on cellulose from water – AN (X_{AN} =0.05 and X_{AN} =0.3) mixtures.

REFERENCES

[1] Klenkova, N.I. Structure and Reactivity of Cellulose; *Nauka:* Leningrad, 1976.
[2] Papkov, S.P. and Fainberg, E.Z. Interaction of Cellulose and Cellulose Materials with Water; *Khimiya:* Moscow, 1976.
[3] Reizin'sh, R.E. Structurization in Suspensions of Cellulose Fibers, Zinatne: Riga, 1987.
[4] Budtov, V.P. Physical Chemistry of Polymer Solutions, *Khimiya:* Moscow, 1992.
[5] Bochek, A.M. in Cellulose and Cellulose Derivatives: Physicochemical Aspects and Industrial Applications, Kennedy, J.F., Phillips, G.O., Williams, P.O., and Picculel, L., Eds., Cambridge: *Woodhead,* 1995, pp. 131-137.
[6] Syrnikov, Yu.P. *Khim. Drev.,* 1988, N 4, pp. 3-9.
[7] Antonchenko, V.Ya. Water Physics, *Naukova Dumka:* Kiev, 1986.
[8] Pelipets, O.V. Investigation of thermodynamics of vaporization and molecular structure of $ErCl_3$, $EuBr_2$, and $EuCl_2$ using high temperature mass-spectrometry and gas electronography, Ph.D. Thesis, Institute of Solution Chemistry of RAS, Ivanovo, 2000.
[9] Smyannikov, P.P.; Basova, T.V.; Grankin, V.M.; Igumenov, I.K. *J. Porph. Phthaloc.* 2000, N 4, pp. 271-277.
[10] Ewing, G.; Sterm, K.H. *J. Phys. Chem.* 1975, vol. 78, pp. 1998-2005.
[11] Sidorov, L.N.; Korobov, M.V.; Zhuravleva, L.V. Investigations of Thermodynamics by Means of Mass-spectrometry, Moscow State University: Moscow, 1985, 208p..
[12] Grebennikova, S.F.; Kynin, A.T.; Negodyaeva, G.S.; Lisyakova, G.V.; Konovalova, L.Y.; Serkov, A.T. *Khim. Volokna* 1988, N 9, pp. 48-49.

[13] Grigoreva, I.S.; Meilihova, E.Z. (eds.), Reference book on Physical Values, *Energoatomizdat,* Moscow, 1991.
[14] Grigoriew, H.; Chmielewski, A.G. *J.Mater. Sci. Lett.* 1997, N 16,1945-1950.
[15] Grigoriew, H.; Chmielewski, A.G.; Amenitsch, H. *Polymer.* 2001, N 42, pp.103-110.
[16] Lebedev, Yu.A.; Miroshnichenko, E.A. Thermochemistry of Vaporization of Organic Substances: Heats of Vaporization and Sublimation and Saturation Vapor Pressure, *Nauka:* Moscow, 1981.
[17] Chickos, J.S.; Acree, W.E. Jr. *J. Phys. Chem. Ref. Data,* 2002, vol. 31, N 2, pp. 537- 698.
[18] A Concise Chemist's Handbook; Nikol'skii, B.P.; Ed.; *Chemistry:* Moscow, 1964; Vol. 3, pp. 237-238.
[19] Zakharov, A.G.; Prusov, A.N.; Voronova, M.I.; et al. *J. Mol. Liq.* 2003, N 103-104, pp. 161-167.
[20] Aguerre, R.J.; Suarez, C.; Viollaz, P.E. *J. Food Sci.* 1986, vol. 51, N 6, pp. 1547-1549.
[21] Telis, V.R.N.; Gabas, A.L.; Menegalli, F.C.; Telis-Romero, *J. Thermochim. Acta,* 2000, vol. 343, pp. 49-56.
[22] Catalan, J.; Diaz, C.; Carcia-Blanco, F. *J. Org. Chem.,* 2001, Vol.66, p.5846-5852.
[23] Lai, J.T.W.; Lau, F.W.; Koga, Y.K.; et al. *J. Solut. Chem.,* 1995, Vol.24, N1, pp.89-102.
[24] Cowie, J.M.G.; Toporowski, P.M.; *Can. J. Chem.,* 1961, Vol.39, pp.2240-2243.
[25] Koga, Y. J.Phys.Chem., 1996, Vol.100, pp.5172-5181.
[26] Koga, Y.; Kasahara, Y.; Yoshino, K.; Nishikawa, K. *J. Solut. Chem.,* 2001, Vol.30, N10, pp.885-893.
[27] 27.Shin, D.n.; Wijnen, J.W.; Engberts, J.B.F.N.; Wakisaka, A. *J. Phys. Chem.*B, 2001, Vol.105, pp.6759-6762.
[28] Shin, D.n.; Wijnen, J.W.; Engberts, J.B.F.N.; Wakisaka A. *J. Phys. Chem.* B, 2002, Vol.106, pp.6014-6020.
[29] 29..Barker, E.S.; Jonas, J. *J. Phys. Chem.,* 1985, Vol.89, pp.1730-1735
[30] Lam, S.Y.; Benoit R.L. *Can. J. Chem.,* 1974, Vol.52, pp.718-722.
[31] Physical Values: A Handbook, Ed. by I. S. Grigor'ev and E. Z. Meilikhov, *Energoatomizdat:* Moscow, 1991, 350 p.
[32] Kipling, J.J.; Tester, D.A. *J. Chem. Soc.,* 1952, pp.4123-4133.
[33] Surov, O.V.; Voronova, M.I.; and Zakharov, A.G. *Zh. Prikl. Khim.,* 2007, Vol. 80, N 1, pp. 43- 48.

[34] O. V. Surov and M. I. Voronova *Russian Journal of Applied Chemistry,* 2007, Vol. 80, No. 12, pp. 2045-2048.
[35] Jayme, G.; Roffael, E. *Papier.,* 1970, Vol. 24, no. 4, pp. 181-186.
[36] Adamson, A.W. Physical Chemistry of Surfaces, New-York: John Wiley & Sons, 1976.
[37] Belousov, V. P.; Morachevskii, A. G.; Panov, M. Yu. Thermal Properties of Solutions of Nonelectrolytes;: *Khimiya:* Leningrad, 1981, 265p.
[38] Herrador, M. A.; Gonzalez, A. G. Talanta, 2002, Vol.56, pp.769-775
[39] Wakisaka, A.; Shimizu, Y.; Nishi, N. et al. *J. Chem. Soc., Faraday Trans,* 1992, Vol. 88, N 8, pp. 1129-1138.
[40] Wakisaka, A.; Abdul-Carime, H.; Yamamoto, Y.; Kiyozumi, Y. *J. Chem. Soc., Faraday Trans,.* 1998, Vol.94, N3, pp. 369-375.
[41] Wakisaka, A.; Takahashi, S.; Nishi, N. *J. Chem. Soc., Faraday Trans.,* 1995, Vol.91, N 22, pp. 4063-4069.
[42] Wei, S.; Shi, Z.; Castleman, A. W. *J. Chem. Phys.,* 1991, Vol. 94, N4, pp.3268-3273.
[43] Zakharov, A. G. ; Voronova, ; Prusov, A. N. et al. Zh. Fiz. Khim, 2006, Vol.80, N8, pp.1472-1478 [Russ. *J. Phys. Chem.,* 2006, Vol. 80, N8, pp.12951299].
[44] Tager, A.A. Physical Chemistry of Polymers, *Khimiya:* Moscow, 1978, 658p.
[45] Dubinin, M.M. ; Astakhov, V.A. *Izv. Akad. Nauk SSSR, Ser. Khim.,* 1971, N 1, pp. 11-16.
[46] Grebennikov, S.F.; Grebennikova, O.D.; Serpinskii, V.V. *Izv. Akad. Nauk SSSR, Ser. Khim.,* 1980, N 2, pp. 453-460.
[47] Koganovskii, A.M.; Klimenko, N.A.; Levchenko, T.M.; Roda, I.G. Adsorption of Organic Substances from Water, *Khimiya:* Leningrad, 1990, 295p.

INDEX

A

absorption, 4, 34, 35, 44, 47, 48, 49
accuracy, 5, 13, 14, 15
acetonitrile, 49
acid, 13, 15, 18
adiabatic, 23
adsorption, 10, 11, 29, 30, 36, 38, 45, 51, 52, 53, 54, 56, 57
adsorption isotherms, 29, 30, 36, 38, 51, 52
agent, 23
air, 12, 18, 55
alkali, 7, 52
amorphous, 2, 3, 51, 52, 54, 58
anomalous, 36, 65
anthracene, 13
aqueous solution, 15, 51, 56, 58
aqueous solutions, 15, 51, 58
aromatic compounds, 13, 51
aromatic hydrocarbons, 2, 59
atmosphere, 52
atoms, 48, 49
availability, 12, 13

B

behavior, 7, 8, 42
biological systems, 23
bipolar, 39
bonding, 36
bonds, 1, 28, 42, 48, 49
Boron, v
by-products, 2, 51

C

calibration, 15, 21, 29, 31, 34
calorimetry, 13, 23
cell, 4, 5, 6, 7, 8, 12, 14, 15, 20, 25
cellulose, i, vi, vii, xi, 1, 3, 4, 5, 6, 8, 9, 10, 11, 12, 16, 17, 18, 19, 20, 21, 23, 25, 29, 30, 31, 34, 35, 36, 37, 38, 39, 43, 44, 45, 46, 47, 48, 49, 51, 52, 54, 55, 56, 57, 59, 64, 65, 66, 67
chemical interaction, 20
chromatography, 13
clusters, 2, 9, 11, 23, 28, 29, 36, 39, 42, 43, 44, 45, 48, 49, 59, 60, 63, 65, 66
CO_2, iv
collisions, 8
combustion, 13
compensation, 18, 20
components, xi, 1, 4, 27, 30, 32, 33, 34, 35, 39, 42, 43, 45, 48, 52, 54, 59, 63
composites, 2, 59
composition, 2, 23, 25, 28, 32, 34, 35, 39, 42, 43, 44, 52, 59, 66
compounds, 2, 8, 13, 51

concentration, 15, 18, 23, 25, 28, 29, 30, 33, 34, 35, 36, 42, 43, 44, 48, 49, 51, 53, 60, 62, 63, 64, 65, 66
condensation, 13, 14, 15, 33, 34
correction factors, 15
correlation, 2, 15, 20, 54, 59
correlation coefficient, 54
correlations, 60
cosine, 8
critical value, 24
cross-sectional, 32
crystalline, 20, 51, 58
crystallinity, 54, 55, 58
crystallites, 2, 51, 57
crystallization, 13
cyclohexanol, 24

D

degradation, 54
degree of crystallinity, 3, 57
dehydration, 36
density, 41
derivatives, 23
desorption, 8, 10, 11, 12, 16, 17, 18, 19, 20, 21, 25, 31, 36, 37, 38
desorption of water, 12, 16, 18, 21
destruction, 43, 48
deviation, 14, 15, 25, 39
diffusion, 1, 3, 54
diffusion rates, 1
dimethylsulfoxide, xi, 27
dipole, 42
direct measure, 26
displacement, 49, 52, 57
distilled water, 12
distribution, 8, 24
diversity, 12

E

effusion, 4, 5, 6, 7, 8, 9, 12, 13, 14, 15, 16, 17, 18, 21, 25, 31, 32, 33, 34, 35
electromagnetic, 12

emulsions, 2, 51
endothermic, 39, 48, 49, 60, 62, 63
energetic characteristics, 38
energy, 54, 58
Enthalpy, 9, 19, 20
entropy, 8, 9, 11, 18, 19, 20, 24, 28, 39, 60, 62, 63
equilibrium, 1, 4, 5, 9, 16, 18, 19, 20, 21, 25, 31, 34, 52
equilibrium sorption, 19, 20
ethers, 1
evacuation, 4, 25
evaporation, 6, 7, 8, 13, 25, 27, 28, 31, 32, 33, 34
experimental condition, 32
express method, 21
extraction, 12, 35

F

fiber, 55, 56, 57
flow, 4, 5, 14, 15, 25, 32
fragmentation, 23
fusion, 10

G

gas, 5, 7, 13, 32, 48, 51, 52, 54, 55, 57, 67
gas phase, 48, 51, 52, 54, 55, 57
gases, 32
Gibbs, 10
glass, 4, 12, 47
grouping, 9
groups, 1, 2, 24, 42, 48, 54, 59

H

heat, 18, 20, 48, 49
heterogeneous, 1
high temperature, 67
humidity, 3, 4, 5, 6, 7, 8, 9, 10, 11, 12, 16, 20
hydration, 3, 20, 66

hydro, 2, 51, 59
hydrogen, 3, 28, 36, 63
hydrogen bonds, 3, 63
hydrolysis, 12
hydrophilic, 51
hydrophobic, 24, 28, 36, 42, 43, 44, 48, 51, 60
hydrophobic properties, 28, 36, 42, 43, 44
hydroxyl, 1
hydroxyl groups, 1

I

immersion, xi, 36, 37, 38
impurities, 13, 20
inactive, 2, 51
integration, 5, 7, 25
interaction, 1, 3, 18, 20, 24, 39, 43, 48, 51, 54, 59, 62, 63
interaction process, 63
interactions, xi, 1, 10, 23, 25, 28, 29, 36, 38, 39, 42, 44, 48, 60, 61, 63, 65, 66
interface, 10, 11, 58
intermolecular, xi, 1, 10, 24, 25, 28, 36, 39, 42, 48
intermolecular interactions, xi, 1, 10, 24, 25, 28, 36, 39, 42, 48
interval, 13, 60, 63
Investigations, 67
ionic, 25
ionization, 8
isolation, 1, 23, 39, 42
isothermal, 51, 59
isotherms, xi, 12, 16, 17, 21, 25, 29, 30, 34, 35, 36, 38, 43, 49, 51, 52, 63, 65

L

Langmuir, 44, 49
law, 8, 25, 42
limitations, 14, 31
linear, 20
liquid nitrogen, 25

liquid water, 3, 4, 9, 10, 11, 24
liquids, 1, 23, 30, 39, 51, 59
low molecular weight, 1, 23, 39, 51, 59

M

macromolecular order, 2, 10, 51
macromolecules, 2, 48, 51, 54
mass loss, 7
mass transfer, 13
maximum sorption, 34, 35
MCC, 55, 56
medical products, 3, 21
melting, 13
metals, 7
microcrystalline cellulose, 52, 58
mixing, 23, 28, 38, 39, 41, 48
mobility, 1
moisture, 3, 16, 18, 47
moisture content, 3, 16, 18, 47
molar volume, 24
mole, 9, 28, 29, 42, 48, 49, 60, 63
molecular beam, 4, 8, 25
molecular mass, 5, 32
molecular structure, 67
molecular weight, 1, 13, 23, 39, 51, 59
molecules, 2, 3, 8, 14, 18, 23, 28, 29, 31, 36, 39, 42, 43, 44, 48, 49, 57, 59, 63, 65, 66
morphological, 1
Moscow, 67, 68, 69
motion, 43, 63, 65

N

naphthalene, 13
natural, 2, 59
network, 42, 63
New York, vii, viii
nitrogen, 4, 25
NMR, 23, 36
normal, 2, 51

O

optimization, 20
organic, xi, 1, 39, 43, 48, 51, 52, 56, 59, 60, 63
organic compounds, 52, 56
organic solvent, 2, 39, 43, 48, 59
organic solvents, 48
Organometallic, iii
orientation, 8
oscillations, 3
oxygen, 36, 48, 49

P

petroleum, 2, 59
petroleum products, 2, 59
pharmaceutical, 23
phenol, xi, 24, 51, 52, 53, 54, 55, 56, 58, 59, 60, 61, 62, 63, 64, 65, 66
physicochemical, 1, 20, 39, 51, 52, 59
physicochemical properties, 1, 39, 52
polymer, 1, 9, 39, 47, 48, 54, 59, 66
polymer chains, 9
polymeric materials, 1, 2, 51
polymerization, 3, 55
polymers, 2, 10, 48, 51, 59
pores, 3, 10, 52, 54
powder, 55, 56
power, 52
prediction, 2, 59
pressure, 3, 5, 6, 7, 8, 9, 11, 12, 14, 15, 16, 17, 18, 21, 25, 26, 31, 32, 33, 34, 42
primary data, 16
pulp, 52
pure water, 27, 28, 33, 36, 43, 48, 60, 63, 65

R

radial distribution, 24
radius, 14, 31
range, 5, 7, 12, 13, 15, 18, 24, 28, 36, 42, 49, 60, 62, 63, 65, 66
RAS, 67
raw material, 52
reactivity, 1
reagents, 1
reciprocal temperature, 18
recrystallization, 13
refractive index, 41
refractometry, 35
relationship, 20
relaxation, 3
resistance, 13, 20, 32
respiratory, 2, 51
retention, 52, 54, 55, 57
rhenium, 5
Russian, 69

S

sample, 4, 5, 6, 7, 8, 11, 12, 16, 17, 18, 20, 29, 34, 35, 47, 52, 53, 55, 58
saturation, 44, 51, 59, 64
sensitivity, 5, 12, 14, 15, 25
shock, 12, 16
sign, 25, 42
simulation, 3
skin, 2, 51
solid phase, 1
solubility, 54, 59, 60, 63
solvation, xi, 2, 24, 28, 38, 59, 63, 64, 65
solvent, 2, 20, 23, 28, 36, 38, 42, 44, 48, 54, 59, 60, 63, 66
solvent molecules, 29, 39
solvents, 2, 12, 21, 39, 42, 47, 59
sorbents, 2, 51, 59
sorption, xi, 1, 3, 9, 10, 12, 16, 17, 18, 19, 20, 31, 34, 35, 36, 39, 43, 44, 48, 49, 51, 52, 54, 58, 59, 63, 64, 65, 66
sorption isotherms, xi, 35, 49, 63
sorption process, 66
species, 28
spectrophotometry, 52
spectroscopy, 23
spectrum, 5, 12, 18, 25

stabilization, 43, 63
stabilize, 43, 48
stainless steel, 4
storage, 20
strong interaction, 28, 60
structure formation, 10
substances, 2, 29, 31, 42, 51
sulfuric acid, 15, 18
supramolecular, 1, 2, 48, 51, 54, 57
surface area, 4
surface energy, 10
surface layer, 48
swelling, 34, 47, 48, 49
synthetic polymers, 2, 59

T

temperature, 3, 5, 8, 12, 13, 16, 18, 20, 25, 26, 32, 33, 34, 65, 67
temperature dependence, 65
thermodynamic, xi, 1, 3, 18, 20, 21, 25, 31, 39, 60, 62
thermodynamic parameters, 18
thermodynamic properties, 39
thermodynamics, 8, 10, 23, 62, 67
thin film, 3
three-dimensional, 24, 42
toluene, 51, 52, 53, 54, 55, 56, 57
toxicity, 23
transfer, 33, 60
transformation, 6, 20
tungsten, 5

U

universal gas constant, 32

V

vacuum, 4, 12, 13, 18, 23
values, 5, 6, 7, 8, 11, 13, 15, 18, 23, 25, 28, 30, 49, 52
vapor, 4, 5, 7, 8, 9, 10, 11, 12, 13, 14, 15, 16, 17, 18, 21, 25, 31, 32, 33, 34, 48, 52
variation, 34, 35
viscosity, 41

W

water absorption, 10
water clusters, 9, 11, 24, 36, 42, 43, 45, 48, 49, 63, 65, 66
water desorption, 8, 9, 10, 11, 15, 16, 18, 19, 20, 21
water evaporation, 5, 8, 28, 36
water sorption, 11, 18, 21, 31, 35, 36, 43, 65
water vapor, 4, 5, 7, 8, 9, 10, 11, 14, 15, 16, 18, 21, 31, 33, 34, 48
weight loss, 13, 25
wetting, 44, 46, 47, 48, 49

X

X-ray diffraction, 23